# XLI Symposium on Nuclear Physics (Cocoyoc 2018)



Journal of Physics: Conference Series Volume 1078

Cocoyoc, Mexico
8-11 January 2018

**Editors:**

**Luis Acosta**
**Paulina Amador-Valenzuela**
**Victor Velazquez**

ISBN: 978-1-5108-7087-1
ISSN: 1742-6588

**Printed from e-media with permission by:**

Curran Associates, Inc.
57 Morehouse Lane
Red Hook, NY 12571

**Some format issues inherent in the e-media version may also appear in this print version.**



Printed by Curran Associates, Inc. (2018)

For permission requests, please contact the Institute of Physics at the address below.

Institute of Physics
Dirac House, Temple Back
Bristol BS1 6BE UK

Phone: 44 1 17 929 7481
Fax: 44 1 17 920 0979

techtracking@iop.org

**Additional copies of this publication are available from:**

Curran Associates, Inc.
57 Morehouse Lane
Red Hook, NY 12571 USA
Phone: 845-758-0400
Fax: 845-758-2633
Email: curran@proceedings.com
Web: www.proceedings.com

# Table of contents

## Volume 1078

2018

**XLI Symposium on Nuclear Physics 2018 (Cocoyoc2018)**
**8–11 January 2018, Cocoyoc, Morelos, México**

**Accepted papers received: 2 August 2018**
**Published online: 30 August 2018**

### Preface



### Papers

doi:10.1088/1742-6596/1078/1/011001

# XLI Symposium on Nuclear Physics

Hacienda Cocoyoc, Morelos, México
January 8-11, 2018

Editors:

Luis Acosta
Instituto de Física
Universidad Nacional Autónoma de México
Paulina Amador-Valenzuela
Instituto Nacional de Investigaciones Nucleares
Víctor Velázquez
Facultad de Ciencias
Universidad Nacional Autónoma de México

## International Advisory Committee

| | |
|---|---|
| Daniel Bardayan | University of Notre Dame, USA |
| Francesco Cappuzzello | INFN-Laboratori Nazionali del Sud, Italia |
| Osvaldo Civitarese | Universidad Nacional de La Plata, Argentina |
| Jerry Draayer | Louisiana State University, USA |
| Alfredo Galindo-Uribarri | Oak Ridge National Laboratory, USA |
| María José García Borge | ISOLDE-CERN, Switzerland |
| Reiner Krücken | TRIUMF, Canada |
| Jorge López | University of Texas at El Paso, USA |
| Takashi Nakamura | Tokyo Institute of Technology, Japan |
| Thomas Papenbrock | University of Tennessee, USA |
| Jorge Piekarewicz | Florida State University, USA |
| Stuart Pittel | University of Delaware, USA |
| Elena Santopinto | Università degli Studi di Genova, Italia |
| W. Michael Snow | Indiana University, USA |
| Adam Szczepaniak | Indiana University, USA |
| Piet Van Isacker | GANIL, France |
| Michael Wiescher | University of Notre Dame, USA |

## Organizing Commmittee

| | |
|---|---|
| Luis Acosta | IF-UNAM |
| Paulina Amador-Valenzuela | ININ |
| Víctor Velázquez | FC-UNAM |

## Sponsors

Instituto de Ciencias Nucleares, UNAM
Instituto de Física, UNAM
Instituto Nacional de Investigaciones Nucleares
División de Física Nuclear de la SMF

IOP Conf. Series: Journal of Physics: Conf. Series **1078** (2018) 011001 doi:10.1088/1742-6596/1078/1/011001

# Photograph of the event

IOP Conf. Series: Journal of Physics: Conf. Series **1078** (2018) 011001 doi:10.1088/1742-6596/1078/1/011001

# Preface

Starting a new decade of meetings, the XLI Symposium on Nuclear Physics was celebrated at the Hacienda Cocoyoc from January 8-11, 2018. The Symposium was conceived from the beginning as a small meeting designed to bring together some of the leading nuclear scientists in the field. Its most distinctive feature is to provide a forum for specialists in different areas of nuclear physics, both theorists and experimentalists, students, postdocs and senior scientists, in a relaxed and informal environment providing them with a unique opportunity to exchange ideas. After the first meeting in Oaxtepec in 1978, the Symposium was organized every year without interruption, arriving now to the 41st edition.

The scientific program consisted of 32 invited talks and 16 posters on a wide variety of hot topics in contemporary nuclear physics, ranging from the traditional fields of nuclear structure and reactions to nuclear astrophysics, hadronic physics, fundamental symmetries, relativistic heavy ions, nuclear instrumentation and applied nuclear physics.

At the meeting there were representatives of many large international collaborations to report on the most recent research progress and the most recent developments in laboratories like Notre Dame, JLab, Cyclotron Institute Texas A&M and Los Alamos (USA); LUNA-LNGS, LNS and LNL from INFN (Italy); TRIUMF and SNOLAB (Canada); IPN-Orsay and IPHC (France); the facilities of RBI (Croatia), iThemba Labs (South Africa) and ELI-NP (Romania), as well as recent results obtained in local Mexican facilities like ININ, LEMA and the Van de Graaff Laboratory Carlos Graef. On the theoretical side, interesting new developments were presented in nuclear dynamics, nuclear structure, Equation of State, neutrino physics, beyond-the-Standard-Model physics, weakly bound nuclei, nuclear clusters, shell model, hadronic physics and low-energy QCD.

The high quality of the talks and posters, the expertise of the speakers and the broad spectrum of subjects covered in the meeting, shows that nuclear physics continues being a very active area at the frontier of scientific research which establishes bridges between many different disciplines, putting together theoretical and experimental researchers and having an extension to the instrumentation developments and applications. Is important to mention the large amount of young researchers and students that participated in this Mexican meeting.

This special volume of Journal of Physics: Conference Series contains 21 contributions that were presented at this year's Symposium.

Finally, we would like to thank all our colleagues for their enthusiastic participation in this 41st edition of the meeting. In particular, we are grateful to the members of the International Advisory Committee for their invaluable suggestions and to the anonymous referees for their time and dedication in helping us to maintain the high standard of the contributions published in the proceedings of the Symposium.

We want as well to thank to N. De Los Ríos for the nice Symposium logo design.

Luis Acosta
Paulina Amador-Valenzuela
Víctor Velázquez

IOP Conf. Series: Journal of Physics: Conf. Series **1078** (2018) 011002 doi:10.1088/1742-6596/1078/1/011002

# Peer review statement

All papers published in this volume of *Journal of Physics: Conference Series* have been peer reviewed through processes administered by the proceedings Editors. Reviews were conducted by expert referees to the professional and scientific standards expected of a proceedings journal published by IOP Publishing.

IOP Conf. Series: Journal of Physics: Conf. Series **1078** (2018) 012001 doi:10.1088/1742-6596/1078/1/012001

# Optical Model description of fusion and elastic scattering of $^{6}$Li + ($^{58}$Ni, $^{64}$Zn) at sub-barrier energies

**E F Aguilera**[1,*] **and F Torabi**[1,2]

[1] Departamento de Aceleradores, Instituto Nacional de Investigaciones Nucleares, Apartado Postal 18-1027, C. P. 11801, México, D. F., México

[2] Department of Physics, Faculty of Science, University of Mazandaran, P.O.Box 47415-416, Babolsar, Iran

E-mail: *`eli.aguilera@inin.gob.mx`

**Abstract.** Recently reported fusion cross sections for $^{6}$Li + ($^{58}$Ni, $^{64}$Zn) have been reduced and compared with each other. Fairly good consistency was found but with some data spread at low energies, possibly due to contributions of direct reactions to the respective inclusive data. This possibility is investigated by doing Optical Model calculations that simultaneously fit the experimental elastic scattering and fusion data. In addition to a realistic bare potential, fusion and direct polarization potentials are introduced which in principle allow to separate the contributions of the corresponding reaction mechanisms. Good fits are obtained for elastic scattering data at all energies but the fusion data for both systems are underpredicted at the lowest energies. These results are consistent with a possibly dominant contribution of direct reactions to the corresponding inclusive data

## 1. Introduction

There has been considerable interest in studying reactions with loosely bound nuclei, stable or radioactive [1, 2]. Among the stable weakly bound nuclei, $^{6}$Li is perhaps the most studied one, with strong current interest [3–12]. Two particular systems having $^{6}$Li as projectile and medium mass targets will be studied in the present work. Fusion data at energies around the barrier were published for the $^{6}$Li + $^{64}$Zn system in 2013 [13] and for the $^{6}$Li + $^{58}$Ni system in 2017 [7]. In the latter reference, the respective barrier parameters were extracted from the São Paulo potential, for both systems, and used to reduce the corresponding data. With this procedure, a direct comparison between the two systems can be done.

The comparison, restricted to the region where the data overlap with each other, showed a fairly good consistency but with some spread of the data at the lower energies (see Fig. 4 of Ref. [7], or the experimental points in Fig. 7 below). This spread may come from some direct channel contributions that could not be separated in the respective inclusive measurements. In the case of $^{6}$Li + $^{58}$Ni, where inclusive proton yields were measured, the Direct Cluster Transfer (DCT) of deuterons or alpha particles to the target may produce protons undistinguishable from the evaporation protons of interest. For example, the transfer of a deuteron to states above the proton emission threshold could lead to protons undistinguishable from the evaporation protons.

Similarly, in the case of $^{6}$Li + $^{64}$Zn the authors concluded that the heavy residues that they measured could possibly include contributions from 1-proton or 1-neutron transfer as well as from DCT of deuterons or alphas. In fact, these authors actually mentioned that such contributions

IOP Conf. Series: Journal of Physics: Conf. Series **1078** (2018) 012001 doi:10.1088/1742-6596/1078/1/012001

may become dominant in the sub-barrier region. One interesting question is whether there is any way of getting information about these direct-channel contributions using only the data that have been published so far.

To shed some light on this issue, it would help to make a simultaneous analysis of the respective elastic scattering and fusion data for each system. By a proper determination of fusion and direct polarization potentials one should be able to obtain fusion cross sections which are consistent with the elastic scattering data and with the dispersion relation. As a matter of fact, such an analysis was recently performed for the $^{6}$Li + $^{58}$Ni system [14]. As a result, it was shown that the fusion excitation function could be reproduced only partially, with discrepancies at the two lowest energies, where the experimental values were underpredicted. This result would be consistent with direct channel contributions in the respective inclusive data. The aim of the present work is to perform the complementary analysis for the $^{6}$Li + $^{64}$Zn system and compare both results.

The optical model extended to describe fusion is briefly described in Sec. 2 while the respective fits to the data are presented in Sec. 3. The experimental and predicted fusion cross sections for the two present systems are compared with each other in Sec. 4 and, finally, the conclusions are given in Sec. 5.

## 2. Extended Optical Model

It's been long known that the parameter-free São Paulo Potential (SPP) [15] can be used as an appropriate bare potential to describe fusion reactions. Crema *et al.* [16, 17], for example, have shown that the SPP gives reliable results in the analysis of fusion reactions with stable weakly-bound projectiles. Among other results, this fact has been used to show that the experimental fusion excitation functions for neutron-halo projectiles are suppressed with respect to the corresponding SPP predictions for energies above the respective Coulomb barriers, while those for proton-halo systems are enhanced [2, 18].

Within the framework of the optical model, fusion has been described as the absorption in a short-range imaginary potential interior to the barrier, $W_{int}$. Such a potential effectively simulates an incoming-wave boundary condition, thus giving a proper barrier penetration model (BPM) prediction. The optical model wave function should correctly describe the respective elastic scattering angular distributions if the fusion cross section calculated this way is to be realistic. So, an appropriate polarization potential should most probably be added to the bare potential. Assumedly, this polarization potential would result from respective couplings to fusion and direct channels. It has been shown for many systems that it is reasonable to decompose the total polarization potential into a direct and a fusion part [19–27]. With this in mind, the following general form for the total optical potential is proposed:

$$U_{TOT} = SPP + V_{Coul} - [iW_{int} + U_F + U_D], \tag{1}$$

where $V_{Coul}$ is the Coulomb potential, $W_{int}$ is a short range Woods-Saxon potential (interior to the barrier: $W_0$ = 50 MeV, $r_0$ = 1. fm, $a_0$ = 0.2 fm), and $U_F, U_D$ are complex polarization potentials that can be associated to couplings to the fusion and direct channels, respectively. $U_F$ ($U_D$) is taken as having a volume (derivative) Woods-Saxon shape, where in each case the radius and difussenes are the same for both, the respective real and imaginary parts. To be more specific, $U_{F,D}$ can be written as

$$U_{F,D} = V_{F,D} + iW_{F,D}, \tag{2}$$

where $V_F$ ($V_D$) and $W_F$ ($W_D$) have identical radius and diffuseness.

Within this extended optical model, the fusion cross section can be obtained from the respective imaginary potentials according to the formula [28]

$$\sigma_{fus} = \frac{2}{\bar{h}v} < \chi^{(+)}|W_{int} + W_F|\chi^{(+)} > \quad , \tag{3}$$

where $\chi^{(+)}$ is the distorted wave function that satisfies the Schrödinger equation with the full optical potential of Eq. 1.

## 3. Simultaneous fit of elastic scattering and fusion data

Elastic scattering measurements were published in 2009 for $^6$Li + $^{58}$Ni [29] at 5 near-barrier energies and also for $^6$Li + $^{64}$Zn [30] at 8 energies. As mentioned in Sect. 1, a full analysis of the first system has been done already [14]. In the following, a simultaneous analysis of elastic [30] and fusion [13] data for the $^6$Li + $^{64}$Zn system will be performed. As a first step, the geometric parameters of the polarization potentials will be fixed. For the fusion polarization potential, the values that have been published in the literature for $^6$Li induced reactions were adopted, that is, a reduced radius of 1.4 fm and a diffuseness of 0.43 fm. These are the same parameter values that were used for $^6$Li + $^{58}$Ni [14].

As for the geometric parameters of the direct polarization potential, the following method was introduced: the elastic scattering data were plotted as a function of the distance of closest approach on a Rutherford trajectory, $d$, according to the expression

$$D = d(A_p^{1/3} + A_t^{1/3}) = \frac{1}{2}D_0\left(1 + \frac{1}{sin(\theta_{c.m.}/2)}\right), \quad with\ D_0 = \frac{Z_p Z_t e^2}{E_{c.m.}}. \tag{4}$$

The corresponding plot is shown in Fig. 1 for the $^6$Li + $^{64}$Zn system, where the data for all measured energies are included. It can be seen that, for low distances, they nicely overlap with each other. The strong absorption distance, $d_S$, corresponding to the point where the ratio $d\sigma/d\sigma_{Ruth}$ is 0.25, is therefore well defined, with a value $d_S$ = 1.6 fm. If one assumes that the falling down of the data with decreasing $d$ is due only to the surface potential $U_D$, it is reasonable to take the radius of this potential as $r_D$ = 1.6 fm, with an appropriate diffuseness to reproduce the respective slope. The dashed curve in Fig. 1 corresponds to a derivative Woods-Saxon shape with radius 1.6 fm. In order to fit the falling-down part of the data, a diffuseness $a_D$ = 0.85 fm is needed. A similar analysis for the $^6$Li + $^{58}$Ni system renders a radius of 1.61 fm and a diffuseness of 0.68 fm [14].

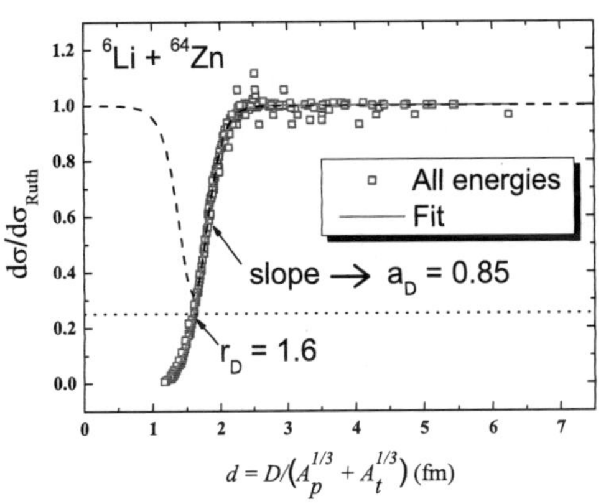


**Figure 1.** Plot of $d\sigma/d\sigma_{Ruth}$ vs $d$ for the data of Ref. [30]. The solid curve corresponds to a three parameter function fit to describe the data while the dashed curve is explained in the text.

The job is then to find the depths of the polarization potentials that best fit both the elastic scattering data and the fusion cross sections. The code FRESCO [31] was used in the present

IOP Conf. Series: Journal of Physics: Conf. Series **1078** (2018) 012001 doi:10.1088/1742-6596/1078/1/012001

work to perform all optical model calculations. After a $\chi^2$ analysis for each of the 8 energies measured, the strengths shown in Fig. 2 were found for the direct part of the potential. The black circles (upper panel) are the imaginary depths while the red ones are the corresponding real depths (lower panel). It is possible to check consistency of these potentials with the dispersion relation [32–34]:

$$V(E) = V(E_S) + \frac{E - E_S}{\pi} P \int_0^{\infty} \frac{W(E')}{(E' - E_S)(E' - E)} dE' \quad . \tag{5}$$

In this expression, $P$ stands for the principal value and $V(E_S)$ is the value of $V(E)$ at a reference energy $E = E_S$.

The imaginary depths were approximated by straight line segments and the red dashed curve is the result of applying the dispersion relation to this approximation. It can be seen that the dispersion relation is well satisfied by these direct polarization potentials.

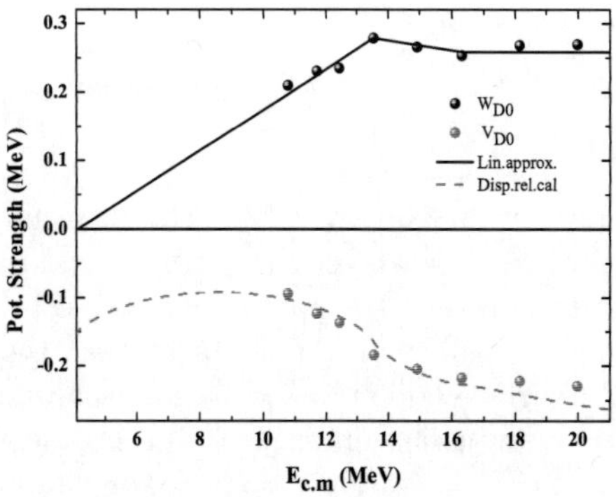


**Figure 2.** Real ($V_{D0}$) and imaginary ($W_{D0}$) parts of the direct polarization potential found for the $^6$Li + $^{64}$Zn system. The dashed curve is the result of applying the dispersion relation to the black curve.

Similarly, the dispersion relation is also satisfied by the fusion part of the polarization potential, as indicated in Fig. 3

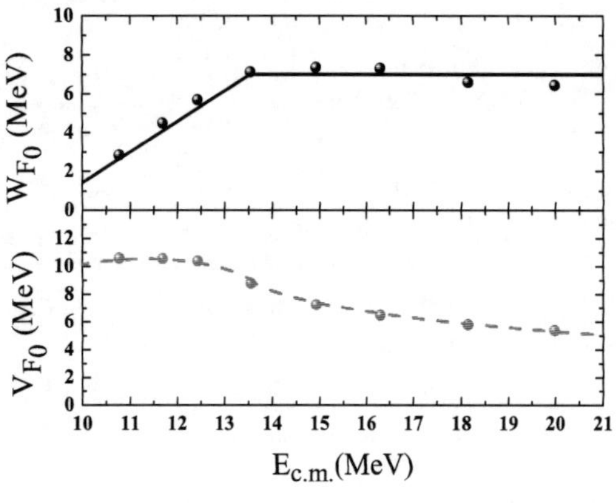


**Figure 3.** Real ($V_{F0}$) and imaginary ($W_{F0}$) parts of the fusion polarization potential found for the $^6$Li + $^{64}$Zn system. The dashed curve is the result of applying the dispersion relation to the black curve.

With the above direct and fusion polarization potentials, the predicted angular distributions are displayed by the curves in Fig. 4, which provide a reasonably good description of the experimental data.

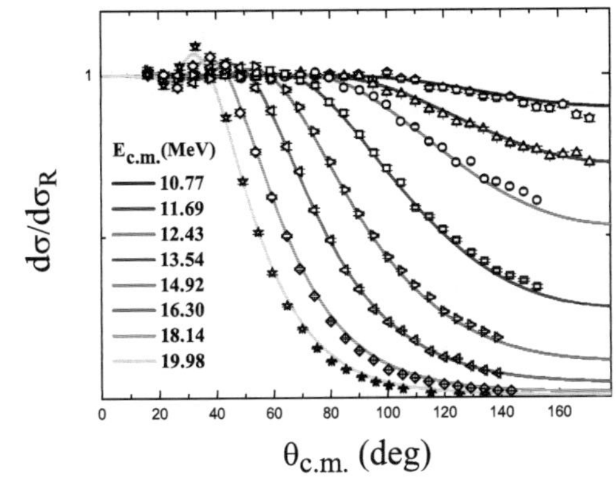


**Figure 4.** Elastic scattering angular distributions for the $^6$Li on $^{64}$Zn system [30] and respective fits obtained in the present work.

The predicted fusion cross sections, calculated from equation 3, are indicated in Fig. 5 by the empty squares, while the circles represent the experimental data. Within the range of energy overlap, the experimental fusion cross sections are well reproduced only for the highest energies, but there is an under prediction for the lowest energies. It is important to mention that the $\chi^2$ analyses at these lower energies showed that no other reasonable potential depths can be found that reproduce the experimental cross sections. Similar to the results for the $^6$Li + $^{58}$Ni system [14], it is speculated that the discrepancies may be explained by a possible direct channel contribution to the respective inclusive measurements. This possibility was considered as quite probable by the authors of the respective experimental paper [13].

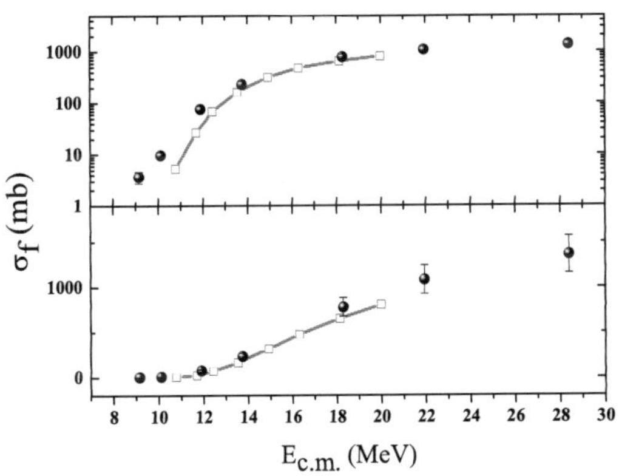


**Figure 5.** Experimental fusion cross sections for the $^6$Li + $^{64}$Zn system [13] (circles) and respective predictions from the present optical model analysis (squares). The upper (lower) panel corresponds to a logarithmic (linear) Y-scale.

## 4. Comparison of both systems

The same type of analysis, performed for the $^6$Li + $^{58}$Ni system [14], rendered the predicted fusion cross sections shown with black diamonds in Fig. 6. The importance of introducing the fusion polarization potential is illustrated by comparing with respective predictions corresponding to $U_F = 0$ (red diamonds). Although including the fusion polarization potential increases considerably the fusion cross sections, there are still discrepancies with the reported data at the two lowest energies. It is possible that such discrepancies may be ascribed to DCT contributions present in the inclusive proton data of Ref. [7].

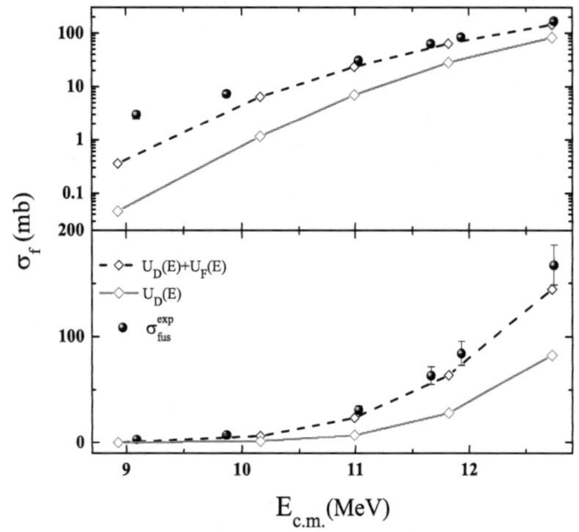


**Figure 6.** Experimental fusion cross sections for the $^6$Li + $^{58}$Ni system [7] (circles) and respective predictions from the optical model analysis [14]. Black diamonds correspond to the full polarization potential, while only the direct part was used for the red diamonds. The upper (lower) panel corresponds to a logarithmic (linear) Y-scale.

Finally, Fig. 7 shows a comparison of reduced cross sections for both systems, including experimental (filled symbols) and predicted (empty symbols) values. The reduction prescription is indicated in the axes labels and UFF stands for the so called Universal Fusion Function UFF $= [2E/(\hbar\omega_0 R_b^2)]\sigma^W$, where $\sigma^W$ represents Wong's function [35]. The barrier parameters used to make the reduction, obtained from the *SPP*, were ($V_b$ = 12.36 MeV, $R_b$ = 9.02 fm, $\hbar\omega_0$ = 3.63 MeV) for $^6$Li + $^{58}$Ni and ($V_b$ = 13.05 MeV, $R_b$ = 9.17 fm, $\hbar\omega_0$ = 3.67 MeV) for $^6$Li + $^{64}$Zn.

It is worth emphasizing that the respective predictions were obtained individually for each energy and for each system. In addition, the respective polarization potentials did always comply with the dispersion relation, consistent with previous findings for many systems [19–22, 26, 27]. It is nice to see that in this reduced plot the theoretical predictions for both systems do actually follow a common path (dashed line), which would be expected for the actual fusion cross sections corresponding to similar systems. This fact adds confidence to the present method of calculating the fusion cross sections. According to these results, assuming that the hypothesis of direct channel contributions in the inclusive data is true, such contributions would be fractionally more significant in the case of $^{64}$Zn than in the case of the $^{58}$Ni target.

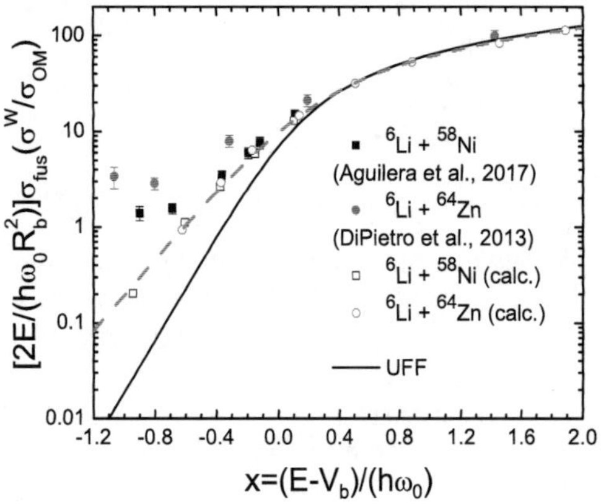


**Figure 7.** Comparison of reduced fusion cross sections for the $^6$Li + ($^{58}$Ni, $^{64}$Zn) systems. Both the experimental (Refs. [7] and [13], respectively) and the predicted values are shown.

## 5. Conclusions

Discrepancies observed in the reduced fusion cross sections for the systems $^{6}$Li + ($^{58}$Ni, $^{64}$Zn) are investigated by performing extended optical model analyses that provide a consistent description of elastic scattering and fusion data. In the present manuscript the $^{6}$Li + $^{64}$Zn system is investigated, complementing the results of a previous work, where an analysis for $^{6}$Li + $^{58}$Ni was done [14].

The SPP was used as a realistic bare potential. Polarization potentials $U_F$ and $U_D$ associated to fusion and direct channel couplings, respectively, were introduced. A method was developed to fix the geometric parameters for $U_D$, while those for $U_F$ were taken from the literature. After fitting simultaneously the elastic scattering and the fusion data, the obtained polarization potentials complied with the dispersion relation. Good descriptions for the elastic scattering angular distributions were obtained. The predicted fusion cross sections agreed with the reported experimental values at the higher energies but discrepancies were observed at low energies.

When compared in a reduced plot, the predictions for both systems agree with each other even though they were derived independently for each system and for each energy. Possible ambiguities in the obtained optical potentials cannot be discarded but, if confirmed, the results would be consistent with a dominant contribution of direct channels in the respective inclusive data at the lowest energies. Exclusive measurements at such energies are desirable.

## Acknowledgments

This work has been partially supported by CONACYT (México) under grant No. CB-01-254619.

## References

[1] Canto L F, Gomes P R S, Donangelo R, Lubian J, Hussein M S 2015 Phys. Rep. **596** 1
[2] Kolata J J, Guimarães V and Aguilera E F 2016 Eur. Phys. J. A **52** 123
[3] Lei J and Moro A M 2017 Phys.Rev. C **95** 044605
[4] Gomez-Camacho A, Lubian J, Zhang H Q, Zhou S-G 2017 Chin. Phys. C **41** 124103
[5] Gomez-Ramos M, Moro A M 2017 Phys. Rev. C **95** 034609
[6] Fisichella M, Shotter A C, Figuera P, Lubian J, Di Pietro A, Fernandez-Garcia J P, Ferreira J L, Lattuada M, Lotti P, Musumarra A, Pellegriti M G, Ruiz C, Scuderi V, Strano E, Torresi D, Zadro M 2017 Phys.Rev. C **95** 034617
[7] Aguilera E F, Martinez-Quiroz E, Amador-Valenzuela P, Lizcano D, García-Flores A, Kolata J J, Roberts A, Rogachev G V, Peaslee G F, Guimarães V, Becchetti F D, Villano A, Ojaruega M, Chen Y, Jiang H, Febbraro M, DeYoung P A, and Belyaeva T L 2017 Phys. Rev. C **96** 024616
[8] Zhang G L, Zhang G X, Hu S P, Yao Y J, Xiang J B, Zhang H Q, Lubian J, Ferreira J L, Paes B, Cardozo E N, Sun H B, Valiente-Dobon J J, Testov D, Goasduff A, John P R, Siciliano M, Galtarossa F, Francesco R, Mengoni D, Bazzacco D, Li E T, Hao X, Qu W W 2018 Phys. Rev. C **97** 014611
[9] Mukeru B, Rampho G J and Lekala M L 2018 J. Phys. G: Nucl. Part. Phys. **45**, 045101
[10] Mukeru B, Rampho G J and Lekala M L 2018 Nucl. Phys. A **969** 60
[11] Kabyshev A M, Kuterbekov K A, Sobolev Yu G, Penionzhkevich Yu E, Kubenova M M, Azhibekov A K, Mukhambetzhan A M, Lukyanov S M, Maslov V A, Kabdrakhimova G D 2018 J. Phys. G: Nucl. Part. Phys. **45** 025103
[12] Diaz-Torres A, Quraishi D 2018 Phys. Rev. C **97** 024611
[13] Di Pietro A, Figuera P, Strano E, Fisichella M, Goryunov O, Lattuada M, Maiolino C, Marchetta C, Milin M, Musumarra A, Ostashko V, Pellegriti M G, Privitera V, Randisi G, Romano L, Santonocito D, Scuderi V, Torresi D, and Zadro M 2013 Phys. Rev. C **87** 064614
[14] Torabi F, Aguilera E F, Ghodsi O N, and Gómez-Camacho A, unpublished
[15] Chamon L C, Carlson B V, Gasques L R, Pereira D, De Conti C, Alvarez M A G, Hussein M S, Cândido Ribeiro M A, Rossi Jr. E S and Silva C P 2002 *Phys. Rev.* C **66** 014610
[16] Crema E, Chamon L C and Gomes P R S 2005 *Phys. Rev.* C **72** 034610
[17] Crema E, Gomes P R S and Chamon L C 2007 *Phys. Rev.* C **75** 037601
[18] Aguilera E F 2016 *J. Phys.: Conf. Series* **730** 012002
[19] Kim B T, Naito M and Udagawa T 1990 *Phys. Lett.* B **237** 19
[20] Kim B T, So W Y, Hong S W and Udagawa T 2002 *Phys. Rev.* C **65** 044616
[21] So W Y, Udagawa T, Kim K S, Hong S W and Kim B T 2007 *Phys. Rev.* C **75** 024610

IOP Conf. Series: Journal of Physics: Conf. Series **1078** (2018) 012001 doi:10.1088/1742-6596/1078/1/012001

[22] Gómez-Camacho A and Aguilera E F 2004 *Rev. Mex. Fis.* **50** Supl. 3, 265
[23] Gómez-Camacho A and Aguilera E F 2004 *Nucl. Phys.* A **735** 425
[24] Gómez-Camacho A, Aguilera E F, Gomes P R S , Lubian J and Padron I 2007 *Nucl. Phys.* A **787** 275c
[25] Gómez-Camacho A, Gomes P R S, Lubian J, Aguilera E F and Padron I 2007 *Phys. Rev.* C **76** 044609
[26] Gómez-Camacho A, Aguilera E F, Lubian J and Gomes P R S 2013 *J. Phys. G: Nucl. Part. Phys.* **40** 035103
[27] Gómez-Camacho A and Aguilera E F 2014 *Phys. Rev.* C **90** 064607
[28] Udagawa T and Tamura T 1984 Phys. Rev. C **29** 1922
[29] Aguilera E F, Martinez-Quiroz E, Lizcano D, Gómez-Camacho A, Kolata J J, Lamm L O, Guimarães V, Lichtenthäler R, Camargo O, Becchetti F D, Jiang H, DeYoung P A, MearsP J, and Belyaeva TL 2009 Phys. Rev. C **79** 021601(R)
[30] Zadro M, Figuera P, Di Pietro A, Amorini F, Fisichella M, Goryunov O, Lattuada M, Maiolino C, Musumarra A, Ostashko V, Papa M, Pellegriti M G, Scuderi V, and Torresi D 2009 *Phys. Rev.* C **80** 064610
[31] Thompson I 1988 *Comp. Phys. Rep.* **7** 167
[32] Satchler G R and Love W 1979 Phys. Rep. **55** 183
[33] Nagarajan M A, Mahaux C C, Satchler G R 1985 Phys. Rev. Lett. **54** 1136
[34] Mahaux C C, Ngô H, and Satchler G R 1986 Nucl. Phys. A **449** 354
[35] C. Y. Wong 1973 Phys. Rev. Lett. **31** 766

IOP Conf. Series: Journal of Physics: Conf. Series **1078** (2018) 012002 doi:10.1088/1742-6596/1078/1/012002

# Recovery of the Center-of-Mass Galilean invariance of a NR-QCD motivated Hamiltonian

**D. A. Amor-Quiroz**[1], **P. O. Hess**[1], **O. Civitarese**[2] **and T. Yepez-Martinez**[1]

[1]Instituto de Ciencias Nucleares, Universidad Nacional Autónoma de México Circuito Exterior, C.U., Apartado postal 70-543, 04510 Mexico City, Mexico.
[2]Departamento de Física, Universidad Nacional de La Plata, C.C.67 (1900), La Plata, Argentina.

E-mail: arturo.amor@nucleares.unam.mx, hess@nucleares.unam.mx, osvaldo.civitarese@fisica.unlp.edu.ar, tochtli.yepez@nucleares.unam.mx

**Abstract.** We evaluate the use of 3D-harmonic oscillator eigenfunctions in the expansion of the fermionic fields of an effective QCD Hamiltonian with the $1/r$ plus a linearly confining potential. Since both the potential and the functions involved in such expansion are non-relativistic, a modified ter-Vehn method is applied to recover the galilean invariance of the center of mass of the mesonic states built with the use of the Random Phase Approximation. Finally, a preliminary calculation for the mesons below 1 $GeV$ is presented.

## 1. Introduction

The Random Phase Approximation (RPA) was introduced in 1953 by Bohm and Pines to describe oscillations in a degenerate electron gas [1]. It consists of a harmonic mapping (bosonization) of fermionic pairs that allows for a correlated ground state, which turns out to be composed of an even number of pairs. The RPA is well-known for preserving (in principle) the symmetries of the total Hamiltonian [2] such as translational and Galilean invariance, though these symmetries can be broken due to some of the following reasons:

*a*) Use of empirical (non-self-consistent) single-particle energies.
*b*) Use of interactions that are not translational invariant.
*c*) Truncation of the configuration space.

In the present document we discuss how to restore the center of mass (CM) symmetries of the RPA calculations applied to the model presented in [3, 4], in which the fermionic fields are expanded in 3D-harmonic oscillator (3D-h.o.) eigenfunctions for solving an effective QCD Hamiltonian with a static potential. Such basis presents many analytical results that simplify intermediate computations [5, 6], though the involved wavefunctions are non-relativistic and thus they prevent the RPA calculations from maintaining the appropriate symmetries.

In [5, 6] the use of Talmi-Moshinsky (TM) transformations was proposed as a tool to identify and remove the spurious excitations originated from frame-dependant dynamics [7, 8]. This could be achieved by restricting the RPA calculations to a subspace of two-particle functions where the bosonized pairs do not present CM excitations. Such procedure has the advantage of

IOP Conf. Series: Journal of Physics: Conf. Series **1078** (2018) 012002 doi:10.1088/1742-6596/1078/1/012002

reducing the size of the involved matrices and therefore requires less computing power. However, this approach may come to be too restrictive as in RPA calculations the CM excitations have to be removed not only from pairs but for any number of fermions. If, for example, there are two pairs in the final state, each pair should be able to move with its own CM but only the CM from the two pairs together has to be substracted, that is, from a 4-particle system.

Consequently, in the present document we prefer to evaluate the use of a modified Meyer-ter-Vehn method [9] which consists of treating the RPA equations in such a way that it is possible to separate the spurious states.

## 2. Brief description of the model

The model presented in [3] consists of an effective QCD Hamiltonian inspired in the Coulomb gauge formalism, where the quarks interact through a static potential, which simulates the effect of the gluons.

In the same way as in QED, implementing the Coulomb gauge gives rise to a local and instantaneous potential. Lattice calculations have shown that this potential is approximately of the form $-\frac{\alpha}{r}+\beta r$ [10], where $r=|\boldsymbol{x}-\boldsymbol{y}|$ is the relative distance between two points of the color charge densities. Such interaction is known in the literature as Cornell potential and its linear component reproduces the phenomenon of confinement, since it makes energetically less favorable to increase the distance between color charges. Because of this, the Cornell potential simulates the effect of gluons, which makes possible the treatment of a simplified theory where the only fields involved are fermionic.

This approach to the problem is equivalent to replacing the Faddeev-Popov operator by the static potential [11]

$$\langle \boldsymbol{x}|\frac{1}{\boldsymbol{\nabla}\cdot\boldsymbol{\mathcal{D}}}(-\boldsymbol{\nabla}^2)\frac{1}{\boldsymbol{\nabla}\cdot\boldsymbol{\mathcal{D}}}|\boldsymbol{y}\rangle \longrightarrow V(|\boldsymbol{x}-\boldsymbol{y}|)=\frac{-\alpha}{|\boldsymbol{x}-\boldsymbol{y}|}+\beta|\boldsymbol{x}-\boldsymbol{y}| \; . \tag{1}$$

Thus, the effective Hamiltonian to be solved in the present model is given by

$$\begin{aligned}
\boldsymbol{H} &= \boldsymbol{H}_{K_q}+\boldsymbol{H}_{m_q}+\boldsymbol{H}^{q-q}_{Coulomb}\\
\boldsymbol{H}_{K_q} &= \int \boldsymbol{\psi}^\dagger(\boldsymbol{x})[-i\boldsymbol{\alpha}\cdot\boldsymbol{\nabla}]\boldsymbol{\psi}(\boldsymbol{x})d\boldsymbol{x}\\
\boldsymbol{H}_{m_q} &= \int \boldsymbol{\psi}^\dagger(\boldsymbol{x})[\beta m_0]\boldsymbol{\psi}(\boldsymbol{x})d\boldsymbol{x}\\
\boldsymbol{H}^{q-q}_{Coulomb} &= -\tfrac{1}{2}\int \rho^{(q)}_{C}(\boldsymbol{x})V(|\boldsymbol{x}-\boldsymbol{y}|)\rho^{(q)C}(\boldsymbol{y})d\boldsymbol{x}d\boldsymbol{y} \; ,
\end{aligned} \tag{2}$$

where $\alpha_i$ and $\beta$ are related to the Dirac matrices $\gamma^\mu$ via $\alpha_i=\gamma^0\gamma^i$ and $\beta=\gamma^0$; while $m_0$ is the bare mass of the quark. For the purpose of this work, the bare mass for the $u$ and $d$ quarks is assumed to be the same though lower than the mass of the quark $s$, *i.e.*,

$$(m_0)_{YT}=m_{u,d}\delta_{Y,\frac{1}{3}}\delta_{T,\frac{1}{2}}+m_s\delta_{Y,-\frac{2}{3}}\delta_{T,0} \; . \tag{3}$$

On the other hand,

$$\rho^{(q)C}=\sum_{c'c}\boldsymbol{\psi}^\dagger{}_{cf}T^C_{cc'}\boldsymbol{\psi}_{c'f} \; , \tag{4}$$

is the color charge density restricted to the quark sector $(q)$ only.

The main characteristic of the model is that the spatial part of the fermionic fields is expanded in the non-relativistic 3D-h.o. basis in the coordinate space [12], *i.e.*, the angular functions are the spherical harmonics while the radial part is given by

$$\begin{aligned} R_{nl}(r) &= \mathcal{N}_{nl} r^l e^{-\frac{\lambda}{2}r^2} L_n^{l+\frac{1}{2}}(\lambda r^2) \\ \mathcal{N}_{nl} &= \left[\frac{2(n!)}{\Gamma(n+l+\frac{3}{2})}\right]^{\frac{1}{2}} \lambda^{\frac{3}{4}+\frac{l}{2}} \\ n &= \frac{N-l}{2} , \end{aligned} \tag{5}$$

where $L_n^\alpha(x)$ are the associated Laguerre polynomials and $\gamma = \frac{1}{\sqrt{\lambda}}$ is the harmonic oscillator lenght. The latter is not assumed as a free parameter, as it is chosen via a variational principle in which the ground state energy of the RPA calculations is minimized.

Afterwards, the TM transformations are introduced to refer the couplings of two-points wavefunctions to the Jacobi coordinates $\boldsymbol{R} = \frac{1}{\sqrt{2}}(\boldsymbol{x}+\boldsymbol{y})$ and $\boldsymbol{r} = \frac{1}{\sqrt{2}}(\boldsymbol{x}-\boldsymbol{y})$

$$\begin{aligned} &[\Psi_{n_a l_a}(\boldsymbol{x}) \otimes \Psi_{n_b l_b}(\boldsymbol{y})]^L_M = \\ &\textstyle\sum_{n_r l_r n_R l_R} \langle n_r l_r, n_R l_R; L | n_a l_a n_b l_b; L\rangle \left[\Psi_{n_r l_r}(\boldsymbol{r}) \otimes \Psi_{n_R l_R}(\boldsymbol{R})\right]^L_M . \end{aligned} \tag{6}$$

Such transformations allow for analytical integration, as the only integrals involved in the matrix elements of the potential are of the form [3]

$$\int_0^\infty r^2 dr R_{n_r' l_r}(r) V(\sqrt{2}r) R_{n_r l_r}(r) \tag{7}$$

while the dependence on the $\boldsymbol{R}$ coordinate sums out. It is to be noted that the definition of the Jacobi coordinates do not coincide with the CM and relative coordinates, though their definitions are closely related for pairs composed of particles with equal masses.

Once the Hamiltonian and the basis are set, the RPA method is applied to diagonalize the effective Hamiltonian while removing the spurious CM excitations as explained in Section 3. Later, an *a posteriori* flavor mixing procedure [13] is set for the $\eta$ and $\eta'$ pseudoscalar mesons. Finally, the resulting calculations are compared with those published in [3].

## 3. The ter-Vehn method

Here we present a generalization of the Meyer-ter-Vehn method which consists in allowing the operators to couple in pairs to a certain total representation.

We start by assuming a particle-hole formulation given by the fermionic operators

$$\begin{aligned} \boldsymbol{b}^\dagger_{a\mu_a} &\equiv \boldsymbol{q}^\dagger_{\frac{1}{2}a\mu_a} \\ \boldsymbol{d}_{a\mu_a} &\equiv \boldsymbol{q}^\dagger_{-\frac{1}{2}a\mu_a} , \end{aligned} \tag{8}$$

where the operators belong to a given representation $a$ with possible magnetic projections $\mu_a$; whereas the index $\pm\frac{1}{2}$ refers to a particle (hole) operator above (below) the Fermi level in such manner that, by the Pauli exclusion principle, any attemp to create a particle below this level will vanish

$$\boldsymbol{q}^\dagger_{-\frac{1}{2}a\mu_a}|0\rangle = \boldsymbol{d}_{a\mu_a}|0\rangle = 0 . \tag{9}$$

The creation and annihilation operators of a particle-hole pair are defined as couplings to a certain total representation $(\Gamma\mu)$ as follows

$$\begin{aligned} \gamma^\dagger_{a\bar{b};\Gamma\mu} &= \sum_{\mu_a\mu_b} \langle a\mu_a, \bar{b}\bar{\mu}_b | \Gamma\mu\rangle \boldsymbol{b}^\dagger_{a\mu_a} \boldsymbol{d}^\dagger_{\bar{b}\bar{\mu}_b} \\ \gamma^{a\bar{b};\Gamma\mu} &\equiv \left(\gamma^\dagger_{a\bar{b};\Gamma\mu}\right)^\dagger , \end{aligned} \tag{10}$$

as the shorthand notation for the Clebsch-Gordan (CG) coefficients refers to the product of an arbitrary number of different yet independent algebras. For example, in the present model the pair operators represent non-colored meson-like states described by the $SU_j(2)\times SU_f(2)\times SU_c(3)$, thus giving the following structure for the CG:

$$\langle a\mu_a, \bar{b}\bar{\mu}_b|\Gamma\mu\rangle = \langle j_a m_a, j_b - m_b|JM_J\rangle\langle T_a T_{z_a}, T_b - T_{z_b}|TT_z\rangle\langle(10)c_a,(01)\bar{c}_b|(00)0\rangle_1 \ . \tag{11}$$

By doing this, one has to be careful in defining the transformation rules for lowering or rising the indices.

$$\begin{aligned}\gamma^{\dagger\bar{a}b;\Gamma\mu} &= \sum_{\mu_a\mu_b}\langle\bar{a}\bar{\mu}_a, b\mu_b|\Gamma\mu\rangle \boldsymbol{b}^{\dagger\bar{a}\bar{\mu}_a}\boldsymbol{d}^{\dagger b\mu_b} = (-1)^{\phi_\mu}\gamma^\dagger_{a\bar{b};\Gamma\bar{\mu}} \\ \gamma_{\bar{a}b;\Gamma\mu} &\equiv \left(\gamma^{\dagger\bar{a}b;\Gamma\mu}\right)^\dagger \ .\end{aligned} \tag{12}$$

The phase $(-1)^{\phi_\mu}$ will be fully determined by the CG involved in the couplings, though they will generically depend on the total magnetic projection $\mu$ and not on its components $\mu_a$ or $\mu_b$ [14].

With the above definitions, the standard RPA equations are given by

$$\sum_{ab}\begin{pmatrix}\mathcal{A}^{(\Gamma\mu)(\Gamma'\mu')}_{(ab)(a'b')} & \mathcal{B}^{(\Gamma\mu)(\Gamma'\mu')}_{(ab)(a'b')} \\ \mathcal{B}^{(\Gamma\mu)(\Gamma'\mu')}_{(ab)(a'b')} & \mathcal{A}^{(\Gamma\mu)(\Gamma'\mu')}_{(ab)(a'b')}\end{pmatrix}\begin{pmatrix}X^\alpha_{ab;\Gamma} \\ Y^\alpha_{ab;\Gamma}\end{pmatrix}\delta_{\Gamma'\Gamma}\delta_{\mu'\mu} = \hbar\Omega^\Gamma_\alpha\begin{pmatrix}X^\alpha_{a'b';\Gamma} \\ -Y^\alpha_{a'b';\Gamma}\end{pmatrix}\delta_{\Gamma'\Gamma}\delta_{\mu'\mu} \ . \tag{13}$$

where $\mathcal{A}$ and $\mathcal{B}$ are the *forward* and *backward* matrices respectively. The ter-Vehn method allows us to remove the couplings between *intrinsic* and spurious states from the Hamiltonian. This is equivalent to solving the same system in equation (13) but modifying the forward and backward matrices as shown below

$$\begin{aligned}\tilde{\mathcal{A}}^{(\Gamma\mu)(\Gamma\mu)}_{(ab)(a'b')} &= \mathcal{A}^{(\Gamma\mu)(\Gamma\mu)}_{(ab)(a'b')} - \frac{1}{2}\left(\mathcal{Z}^{(R)}_{(ab)(a'b')} + \mathcal{Z}^{(P)}_{(ab)(a'b')}\right) \\ \tilde{\mathcal{B}}^{(\Gamma\mu)(\Gamma\mu)}_{(ab)(a'b')} &= \mathcal{B}^{(\Gamma\mu)(\Gamma\mu)}_{(ab)(a'b')} - \frac{1}{2}\left(\mathcal{Z}^{(R)}_{(ab)(a'b')} - \mathcal{Z}^{(P)}_{(ab)(a'b')}\right) \ .\end{aligned} \tag{14}$$

The structure of the $\mathcal{Z}$ coefficients is given in [17].

Such method does not remove the spurious state corresponding to pure oscillations of the CM coordinate. However, this mode is approximately decoupled and appears at a finite energy. In the particular case of the present model, the energy of the pure CM state turns to be of the order of 3 GeV and thus it lies far above the mesonic states evaluated by this method (*i.e.*, below 1 GeV) [17].

## 4. Results

An exploratory spectrum of the mesons below 1 GeV was already shown in [3], though the CM contribution was not eliminated there and thus compromises the self-consistency of the found solutions. However, such document can be regarded as a feasibility test for further calculations.

Here we present some preliminary calculations that were obtained by implementing the ter-Vehn method to the same fit used in [3], whose values are given in Table 1. This has the purpose of evaluating the CM contributions to the previous results.

The fit in [3] was chosen to reproduce the mass of the pion-like states, while the flavor mixing interaction $H_{FM}$ was selected to reproduce as best as possible the $\eta'$ meson.

IOP Conf. Series: Journal of Physics: Conf. Series **1078** (2018) 012002 doi:10.1088/1742-6596/1078/1/012002

| $m_{u,d}[GeV]$ | $m_s[GeV]$ | $\alpha$ | $\beta[GeV^2]$ | $H_{FM}$[GeV] |
|---|---|---|---|---|
| 0.05 | 0.31 | 0.16 | 0.40 | 0.170 |

**Table 1.** Set of parameters for performing the RPA calculations.

| Meson-like state | Exp. value | RPA (ref. [3]) | RPA (CM-removed) |
|---|---|---|---|
| $\pi$ | 139 | 136 | 144 |
| K | 495 | 465 | 476 |
| $\eta$ | 547 | 356 | 363 |
| $\eta'$ | 957 | 958 | 963 |
| $\rho$ | 770 | 676 | 679 |
| $\omega$ | 782 | 676 | 694 |
| $K^*$ | 892 | 942 | 941 |
| $\phi$ | 1020 | 1145 | 1200 |

**Table 2.** Experimental and RPA masses [MeV] for pseudoscalar and vector mesons for the Set of parameters from Table 1 after having removed the CM via the ter-Vehn method and implementing an *a posteriori* flavor mixing procedure for the $\eta$ and $\eta'$ mesons.

It can be observed from Table 2 that the CM motion may be neglected for states such as the $K^*$, though its contribution can reach up to 55 MeV for the $\phi$ meson. Nevertheless, the most noticeable feature of the CM removal procedure is that it breaks the degeneracy from the $\rho$ and $\omega$ states by rising the latter as it should. This is due to a small flavor mixing allowed by the structure of the vacuum in the ter-Vehn method, even if the Hamiltonian preserves the flavor symmetry.

## 5. Conclusions

The results shown in Table 2 exhibit that even without removing the spurious states coming from the CM motion, a fairly reasonable reproduction of the mesons below 1 GeV can be achieved. Nonetheless, the involved physics are improved by the ter-Vehn method and thus, a refitting of the whole parameters is strongly encouraged.

The physics of the model can also be improved, for example, by introducing a dynamical chiral symmetry breaking or by adding spin-spin interactions. Subsequently, the model can be tested over multiquark systems [15] and by computing decaying widths [16].

The advantages obtained up to the current state of the model are summarized below:

- It is a simple yet semi-realistic model. Even though the spectra can be improved, its current state is relatively good.
- The harmonic oscillator basis allows analytic integration, while the integrals involved in the interaction matrix elements are well behaved.
- The ter-Vehn method restores the Galilean and traslational symmetries broken due to the selection of the basis.
- Decoupling the spurious states gives rise to a small flavor mixing evident in the $\omega$-$\rho$ splitting.

Because of the above, the conclusion is that the model is feasible but requires further research to be able to compete with the current approaches for non-perturbative QCD.

Details of the present calculations will be given in [17].

IOP Conf. Series: Journal of Physics: Conf. Series **1078** (2018) 012002 doi:10.1088/1742-6596/1078/1/012002

**Acknowledgements**
One of the authors (T.Y-M) thanks the National Research Council of Argentina (CONICET) for a post-doctoral scholarships. (O.C.) is a member of the scientific career of the CONICET. (P.O.H and D.A.A-Q) acknowledges financial help from DGAPA-PAPIIT (IN100418) and from CONACYT (Mexico, grant 251817). This work has been financially supported by the CONICET (PIP-282) and by the ANPCYT. D.A.A-Q. acknowledges a scholarship received from CONACyT.

## References

[1] D. Bohm and D. Pines, Phys. Rev. 92 (3), **609**, (1953).
[2] D. J. Thouless, Nucl. Phys. **22**. (1961) 78.
[3] A. Amor-Quiroz, T. Yépez-Martínez *et al*, Int. J. Mod. Phys, **26** (2017) 1750082.
[4] T. Yépez-Martínez, D. A. Amor-Quiroz *et al*, J. Phys., Conf. Ser. **876** (2017) 012022.
[5] P. Bicudo, M. Cardoso *et al*, Phys. Rev. D **94**, (2016) 054006.
[6] A. Amor-Quiroz, T. Yépez-Martínez *et al*, J. Phys., Conf. Ser. **912** (2017) 012028.
[7] Gintautas P. Kamuntavičius, J. of Math. Phys., **55** (2014) 042103.
[8] Algirdas Deveikis and Gintautas P. Kamuntavičius J. of Mod. Phys., **6** (2015).
[9] J. Meyer-ter-Vehn, Z. Physik. A **289**, (1979).
[10] K.J. Juge, J. Kuti, and C.J. Morningstar, Nucl. Phys. Proc. Suppl. **63** (1998) 326.
[11] A. P. Szczepaniak and E. S. Swanson, Phys. Rev. D **65** (2001) 025012.
[12] T. Yépez-Martínez, P. O. Hess, Phys. Rev. C **81** (2010) 045204.
[13] A. De Rújula, H. Georgi and S. L. Glashow, Phys. Rev. D **12** (1975) 147.
[14] J. Escher and J. P. Draayer, J. of Math. Phys. **39** (1998) 5123.
[15] T. Yépez-Martínez, O. Civitarese and P. O. Hess, Eur. Phys. J. A **54** (2018) 22.
[16] T. Yépez-Martínez, O. Civitarese and P. O. Hess, Int. J. Mod. Phys, **27** (2018) 1850001.
[17] A. Amor-Quiroz, O. Civitarese *et al* Int. J. Mod. Phys, *to be published.*

IOP Conf. Series: Journal of Physics: Conf. Series **1078** (2018) 012003 doi:10.1088/1742-6596/1078/1/012003

# Effect of resonance states of the weakly bound $^{6}$Li on total fusion in reactions with targets $^{28}$Si, $^{96}$Zr and $^{209}$Bi

**A. Gómez Camacho**

Departamento del Acelerador, Instituto Nacional de Investigaciones Nucleares, Apartado Postal 18-1027, C.P. 11801, México, D.F.

E-mail: arturo.gomez@inin.gob.mx

**Abstract.** Continuum Discretized Coupled-Channel calculations of total fusion cross sections for reactions induced by the weakly bound nucleus $^{6}$Li with targets $^{28}$Si, $^{96}$Zr and $^{209}$Bi at energies around the Coulomb barrier are presented. In the cluster structure frame of $^{6}\text{Li}\rightarrow \alpha+d$, short-range absorption potentials are considered for the interactions between the $\alpha$- and $d$-fragments with the targets. In the model, the effect of resonance ($l = 2, J^{\pi} = 3^{+}, 2^{+}, 1^{+}$) and non-resonance breakup states of $^{6}$Li on fusion is studied by using two approaches: (1) by omitting the resonance states from the full discretized CDCC breakup space and 2) by considering only the resonance discretized sub-space. Among other things, it is found that resonance breakup states produce strong repulsive polarization potentials that lead to fusion suppression. The particular effect on fusion of a single resonance is calculated. It is found that resonances $2^{+}$ and $1^{+}$ have a more significant role on incomplete fusion, while the $3^{+}$ on complete fusion.

## 1. Introduction

One of the most interesting and important topics of research of reactions with weakly bound nuclei is to understand the effects of breakup on elastic scattering and fusion processes. It is already well accepted that couplings to continuum breakup states of weakly bound projectiles have a very strong effect on these nuclear mechanisms. As a matter of fact, consideration of breakup is essential to fit the experimental data [1, 2, 3, 4, 5]. The characteristic low binding energy associated to weakly bound projectiles can affect reaction processes mainly in two ways. Namely, the static effect due to the large diffusivity matter density and the high breakup probability. The stretched density lowers the Coulomb barrier and hence enhances fusion. On the other hand, the high breakup probability produces strong repulsive couplings that raise the barrier and then suppresses fusion. Also, reactions with weakly bound projectiles present characteristic nuclear processes like complete (CF) and incomplete (ICF) fusion. Complete fusion can proceed via a direct (DCF) or sequential (SCF) processes. DCF is a mechanism similar to fusion between strongly bound nuclei, that is, fusion that occurs without a previous excitation of breakup channels. SCF takes place when all projectile fragments fuse to the target after a previous breakup. ICF occurs when only some fragments are absorbed by the target, while the others fly away to the continuum. Another important process is the elastic breakup (EB), that is, when none of the fragments is captured by the target.

IOP Conf. Series: Journal of Physics: Conf. Series **1078** (2018) 012003 doi:10.1088/1742-6596/1078/1/012003

From an experimental viewpoint, DCF and SCF are very difficult to distinguish, since the evaporation products of these processes are very similar. On the other hand, only until recently, it has been possible to perform separate measurements of complete and incomplete fusion. These measurements have been restricted to reactions with light weakly bound projectiles on medium and heavy mass targets. In this case, the compound nucleus decays by emission of uncharged particles, mainly neutrons and $\alpha$-particles [6, 7, 8], thus separate measurements of CF and ICF cross sections can be achieved [9, 10]. In contrast, for light targets, thermal evaporation occurs by an important contribution of charged particles and measurements become more difficult. Therefore, most of fusion measurements for reactions with weakly bound projectiles, correspond to Total Fusion (TF), which is the sum of complete and incomplete fusion. From an experimental viewpoint, DCF and SCF are very difficult to distinguish, since the evaporation products of these processes are very similar. On the other hand, only until recently, it has been possible to perform separate measurements of complete and incomplete fusion. These measurements have been restricted to reactions with light weakly bound projectiles on medium and heavy mass targets. In this case, the compound nucleus decays by emission of uncharged particles, mainly neutrons and $\alpha$-particles [6, 7, 8], thus separate measurements of CF and ICF cross sections can be achieved [9, 10]. In contrast, for light targets, thermal evaporation occurs by an important contribution of charged particles and measurements become more difficult. Therefore, most of fusion measurements for reactions with weakly bound projectiles, correspond to Total Fusion (TF), which is the sum of complete and incomplete fusion.

From a theoretical standpoint, for a complete description of reactions with weakly bound nuclei all of the aforementioned reaction mechanisms should be included in a single calculation. The CDCC (Continuum Discretized Coupled-Channel) formalism [11, 12, 13] is the most widely applied theoretical tool to study breakup states in the continuum. Likewise, the effects of continuum couplings on processes like elastic scattering and fusion can be studied. In fact, the CDCC model has been successfully applied in many nuclear systems [14, 15, 16, 17, 18, 19], since it provides a good description for the observed NCBU, elastic and TF cross sections [14, 15, 20, 21, 22]. Notwithstanding, most CDCC calculations have a shortcoming [21, 23], as they cannot calculate the ICF and CF cross sections separately. That is, a simultaneous description of complete fusion and incomplete fusion is very difficult to achieve. However, recently some efforts have been made in which the simultaneous CDCC calculation of CF and ICF has been carried out. For instance, reactions between the weakly bound projectiles $^{6,7}$Li with targets $^{209}$Bi and $^{198}$Pt [24] and $^{6}$Li with $^{144}$Sm and $^{154}$Sm [25] have been studied. In these calculations, separate incomplete fusion of the fragments $\alpha$-$d$ of $^{6}$Li or $\alpha$-$t$ of $^{7}$L are determined. The technique used in these works, consists of introducing two short-range Woods-Saxon potentials to account for absorption between the fragments, say $a$ and $b$, and the target, while CF is calculated with a different potential in the incident channel. Subsequently, two types of calculations are performed, first, one of the fragment-target potentials, say $W_a$, is turned-off so as to calculate fusion capture of the other fragment $b$ and the incident channel. In a second calculation, this process is repeated for the other fragment, that is, $W_b$ is turned-off, so fusion of fragment $a$ and the incident channel is determined. The contribution to incomplete fusion corresponding to any fragment is then determined by subtracting these results from total fusion. However, there is a serious drawback in this treatment as pointed out in Ref. [26], that is, the competition and correlation between both fragment-target capture processes have been neglected. That is, the capture of a given fragment affects the absorption of the other. Besides, it is assumed in Ref. [24] that breakup of $^{6}$Li happens mainly in the $\alpha - d$ channel, however, recent experimental data show that other important breakup triggered by neutron transfer and deuteron pick processes are significantly important [27]. The effect of neglecting the competition and correlation between the fragment capture absorption processes has been discussed in Ref. [26] for the $^{6}$Li+$^{209}$Bi system. More realistic approaches have been proposed to separately calculate CF and ICF, for instance, the classical dynamical reaction model described in Refs. [28, 29, 30, 31]. This model unambiguously calculates individual contributions to incomplete fusion from the fragments, besides the contributions to CF from sequential and direct processes are explicitely determined. Among other things, it is shown in this model that sequential complete fusion has a more important role than previously assumed. However, this model being a classical model, can not treat fusion for energies below the Coulomb barrier, where tunneling effects are essential.

The CDCC model has been used to calculate total fusion cross sections of several weakly bound systems. For instance, it has been applied for reactions between the projectiles $^{6,7}$Li with targets $^{59}$Co and $^{209}$Bi [16], or the nucleus $^{11}$Be on the target $^{208}$Pb [15]. One of the most important aspects studied in Refs. [16, 15], is to clarify whether breakup produces enhancement or suppression effects on total fusion, as well as its energy dependence. It may be mentioned that for the projectile $^{11}$Be, total fusion is calculated in Ref. [15] by the absorption of the its center of mass. That is, since breakup occurs mainly by the process $^{11}Be \rightarrow ^{10} Be + n$, thus the capture of the center of mass of $^{11}$Be implies the absorption of $^{10}Be$, which carries most of the mass and charge of the projectile. However, this is not the case for nuclei such as $^{6,7}$Li, for which the main breakup channels are $^6Li \rightarrow \alpha + d$ and $^7Li \rightarrow \alpha + t$. In both cases the fragment masses are not so different, thus fusion can not be calculated by the absorption of their center of mass. For this reason, two short-range absorption potentials are used to account for capture of $\alpha$ and $d$ for $^6$Li or $\alpha$ and $t$ for $^7$Li as done in Ref. [16]. In this way, complete fusion (both direct and sequential) is accounted for when both fragments are captured, that is, both are inside the region of absorption. Similarly, incomplete fusion is taken into account when only one fragment is inside its respective short-range potential while the other is outside and flies to the continuum. Clearly, this procedure can not distinguish whether CF occurs by a direct or sequential process.

Hence, in the present work, we present a systematic CDCC calculations of total fusion for the weakly bound projectile $^6$Li with targets $^{28}$Si, $^{96}$Zr and $^{209}$Bi at energies around the Coulomb barrier. Following Ref. [16], two short-range imaginary volume Woods-Saxon potentials will be used to account for fusion absorption of the fragments $\alpha$ and $d$ by the target. Besides determining the effect of breakup on fusion, the effects of resonance and nonresonance scattering excited states $^6$Li are studied by extracting the corresponding sub-space from the full discretized space. An interpretation of these effects will be given in terms of the polarization potentials that result from the corresponding couplings. An analysis of effects on fusion from resonance and nonresonance states as function of the target mass is also presented. In fact, so as to isolate the effects of the resonances, couplings to excited states of the target will be ignored.

It is important to mention that this work is a sequence of our previous CDCC calculations of elastic scattering angular distributions and the effect of resonance and nonresonance breakup continuum states of $^6$Li on targets $^{28}$Si, $^{58}$Ni [32] and $^{144}$Sm [33]. In these calculations systematic Woods-Saxon volume and surface potentials were used for the interactions between the $\alpha$-particle and *deuteron* with the target. The effect of resonance and nonresonance couplings was determined by following two different approaches. First, by omitting couplings to/from resonance states from the full discretized CDCC breakup space, secondly by considering only couplings to/from the resonance sub-space. For the calculations with resonance couplings (nonresonance space excluded), the elastic scattering angular distributions were very much similar to that with the full discretized energy space. Therefore, consideration of these states is essential. On the other hand, when resonance states are excluded (nonresonance couplings included) a larger deviation is present, particularly for $^{58}$Ni and $^{144}$Sm at energies below the barrier and for backward dispersion angles. The effects of resonance and nonresonance sub-spaces were explained in terms of the polarization potentials that appear from the corresponding couplings. It was determined that the polarization potentials produced by resonance couplings are strongly repulsive, therefore suppress fusion. Those from the nonresonance sub-space are weakly repulsive and thus have a smaller effect. A comparison regarding the effects of resonance and nonresonance couplings on elastic scattering and those presented in this work for total fusion will be also discussed.

The recent work of Ref. [26] for the $^6$Li+$^{209}$Bi system establishes the significant importance of $2^+$ and $1^+$ resonance states of $^6$Li on ICF and NCBU processes. This is explained in terms of the shorter half-life of these states respect to the $3^+$ resonance. For the $3^+$-resonance having a longer half-life, delayed breakup takes place. Thus, breakup of the $3^+$-state can occur far from the target, at the outgoing part of the trajectory, thus a small effect on ICF is expected. If breakup happens near the target, then it has a more important effect on CF than on ICF. On the other hand, for the $2^+$ and $1^+$ states prompt breakup takes place thus a larger effect on ICF is expected. In a final calculation, we study the effect on fusion of a given resonance by omitting couplings to only this state, then a discussion of the results is presented.

As a last comment, it is necessary to emphasize that our calculations are restricted to the assumption that the main breakup mechanism of $^6$Li occurs into the $\alpha$ and $d$ channel. Recent measurements [27] have established that significant breakup of $^6$Li is triggered by neutron transfer leading to sequential $p-\alpha$ breakup, or via $d$-pickup by $^6$Li leading to sequential breakup of $^8B \rightarrow \alpha+\alpha$. However, these measurements also show that still the main breakup dissociation of $^6$Li occurs through the $\alpha - d$ channel. Treatment of more complicated four-body CDCC calculations of sequential breakup triggered by particle transfer is of current theoretical interest but not attempted in this work.

In section II, a very brief description of the CDCC model is presented as well as the numerical calculations. In section III, a summary and conclusions are presented.

## 2. Brief CDCC description and calculations

A detailed description of the CDCC formalism is given in Refs.[11, 12, 13]. Also, a formal description of how the CDCC is used to study the effects of resonance and nonresonance states is presented in Refs. [32, 33]. In this section, only the basic equations that are required to perform our calculations are presented. The two-body cluster model structure of $^6$Li ($\alpha$-$d$) with ground state energy $E_{g.s.} = -1.47$ MeV is assumed. The model space for constructing discrete breakup states is that described in Ref. [16]. The total wave function of the three-body system ($\alpha$-$d$-$target$) is given by,

$$\Psi(\mathbf{R},\mathbf{r}) = \sum_{\beta} F_\beta(\mathbf{R})\psi^P_\beta(\mathbf{r})\Phi^T_0, \tag{1}$$

where $\Phi^T_0$ is the ground state of the target. $F_\beta(\mathbf{R})$ represents the projectile-target relative radial motion, where the projectile is in the discrete breakup state $\beta$. Here, $\mathbf{R}$ is the relative distance between the target and the center of mass of the cluster $^6$Li. The $\beta$-channel can be the elastic incident channel or any discrete excited state of the projectile. $\psi^P_\beta(\mathbf{r})$ describes the radial relative motion of the ($\alpha$,$d$)-system, where $\mathbf{r}$ is the relative distance between the fragments. The coupled-channel equations are obtained by projecting the equation of motion $\hat{H}\Psi = E\Psi$ onto the discrete radial wave function $\psi^P_\beta(\mathbf{r})$ of the projectile.

$$\begin{aligned}&[\hat{T}_{R,K} + U^{(J)}_{\beta,\beta}(R) - (E - \epsilon_0 - \varepsilon_\beta)]F^{(J)}_\beta(R)\\ &\qquad = -\sum_{\beta'\neq\beta} U^{(J)}_{\beta,\beta'}(R)F^{(J)}_{\beta'}(R).\end{aligned} \tag{2}$$

Where, $\epsilon_0$ is the ground state energy of the target (excited states are not considered) and $\varepsilon_\beta$ the ground and discrete breakup states of the projectile. $U^{(J)}_{\beta,\beta'}$ are the diagonal ($\beta = \beta'$) and non-diagonal ($\beta \neq \beta'$) coupling matrix elements given by,

$$U^{(J)}_{\beta\beta'}(R) =< u_\beta \left| \hat{V}_{dT}(\mathbf{r}_{dT}) + \hat{V}_{\alpha T}(\mathbf{r}_{\alpha T}) \right| u_{\beta'} >, \tag{3}$$

where $u_\beta(r)$ are the normalized square-integrable wave functions known as bin states [11, 32, 33], which are functions of the relative distance $\mathbf{r}$ between the fragments and that satisfy $\mathbf{r_{\alpha T}} = \mathbf{R}+\frac{2}{6}\mathbf{r}$ and $\mathbf{r_{dT}} = \mathbf{R} - \frac{4}{6}\mathbf{r}$. $\hat{V}_{\alpha T}$ and $\hat{V}_{dT}$ are the $\alpha - T$ and $d - T$ real interactions that include Coulomb and nuclear parts and depend on the distance between the fragments to the target. To account for fusion, two short-range imaginary potentials $\hat{W}_{\alpha T}$ and $\hat{W}_{dT}$ will be used for the $\alpha - T$ and $d - T$ interactions, i.e.,

$$W(r_i) = W_0 f^{-1}(r_i), \quad f(r_i) = 1 + exp[(r_i - R_0)/a_0], \tag{4}$$

where $r_i = r_{dT}$ and $r_{\alpha T}$ are distances between the fragments to the target, $a_0$ is the diffuseness and $R_0 = r_0(A_T^{1/3} + A_{\alpha,d}^{1/3})$, $A_T$, $A_\alpha$ and $A_d$ being the target, $\alpha$-particle and deuteron mass numbers respectively. These potentials are equivalent to the use of an incoming boundary condition inside the Coulomb barrier. Total fusion is calculated by,

$$\sigma_{TF} = \frac{2}{\hbar v} \sum_{J,\beta} \frac{1}{2J+1} < F^{(J)}_\beta(R) \left| W^{(J)}_{\beta\beta}(R) \right| F^{(J)}_\beta(R) >, \tag{5}$$

where $v$ is the projectile-target collision velocity in the center of mass and $W^{(J)}_{\beta,\beta}(R) =< u_\beta \left| \hat{W}_{\alpha T}(\mathbf{r_{\alpha T}}) + \hat{W}_{dT}(\mathbf{r_{dT}}) \right| u_\beta >$.

The procedure to construct bin discretized states of $^6$Li is essentially the same as that used in the CDCC calculations of elastic scattering of Refs. [32, 33]. That is, the ground (binding energy -1.47 MeV), resonance and nonresonance bin scattering states of $^6Li \to d+\alpha$ are generated from the Hamiltonian $\hat{h}_{\alpha d}(\mathbf{r}, \mathbf{k})$, using the $\alpha-d$ nuclear and spin-orbit interactions of Ref. [16]. Careful attention is paid to construct resonance states $l = 2$, $j^\pi = 3^+, 2^+, 1^+$ with centroid energies and widths close to the experimental values and to avoid double-counting. The discretization procedure to obtain convergent calculations is carried out as follows: bin states are constructed from an initial energy $\varepsilon_0 = 0$ MeV (above the threshold energy $E_{thres} = -1.47$ MeV) up to a maximum energy $\varepsilon_{\max} = 6.8$ MeV for energies below and around the barrier and $\varepsilon_{\max} = 8$ MeV for higher energies. The relative angular momentum $l$ between the fragments $\alpha$ and $d$ varies from 0 to 3, higher values do not have an effect on the convergent calculations. Partial waves of the relative angular momentum between the center of mass of projectile and of the target are taken up to $L = 60$, while potential multipoles of nuclear and Coulomb interactions are considered up to $\mathcal{K} = 4$. Following Ref. [16], continuum (nonresonant and resonant) breakup states are discretized in equally spaced momentum bins with respect to the relative wave number $k$ between the fragments. Nonresonance bin states $(l = 0, J^+ = 1)$, $(l = 1, J^- = 0, 1, 2)$, $(l = 3, J^- = 2, 3, 4)$ and resonance states $l = 2, J^+ = 3^+, 2^+, 1^+$ are discretized with the same densities as reported in Ref. [16].

The calculations of total fusion cross sections are performed with the use of the FRESCO code described in Ref. [34]. It is important to point out that a similar systematic analysis was reported in Refs. [32, 33], for elastic scattering angular distributions of $^6$Li with targets $^{28}$Si, $^{58}$Ni and $^{144}$Sm. However in these works, global nuclear and imaginary absorption potentials [35, 36, 37] were used for the $d - T$ and $\alpha - T$ interactions. In the present calculations, use is made of the same real nuclear interactions, however to account for fusion absorption, short-range volume Woods-Saxon potentials of Eq.(4), are used for the fragment-target interactions. In the calculations here presented, the parameters of these potentials will be set to, reduced radius

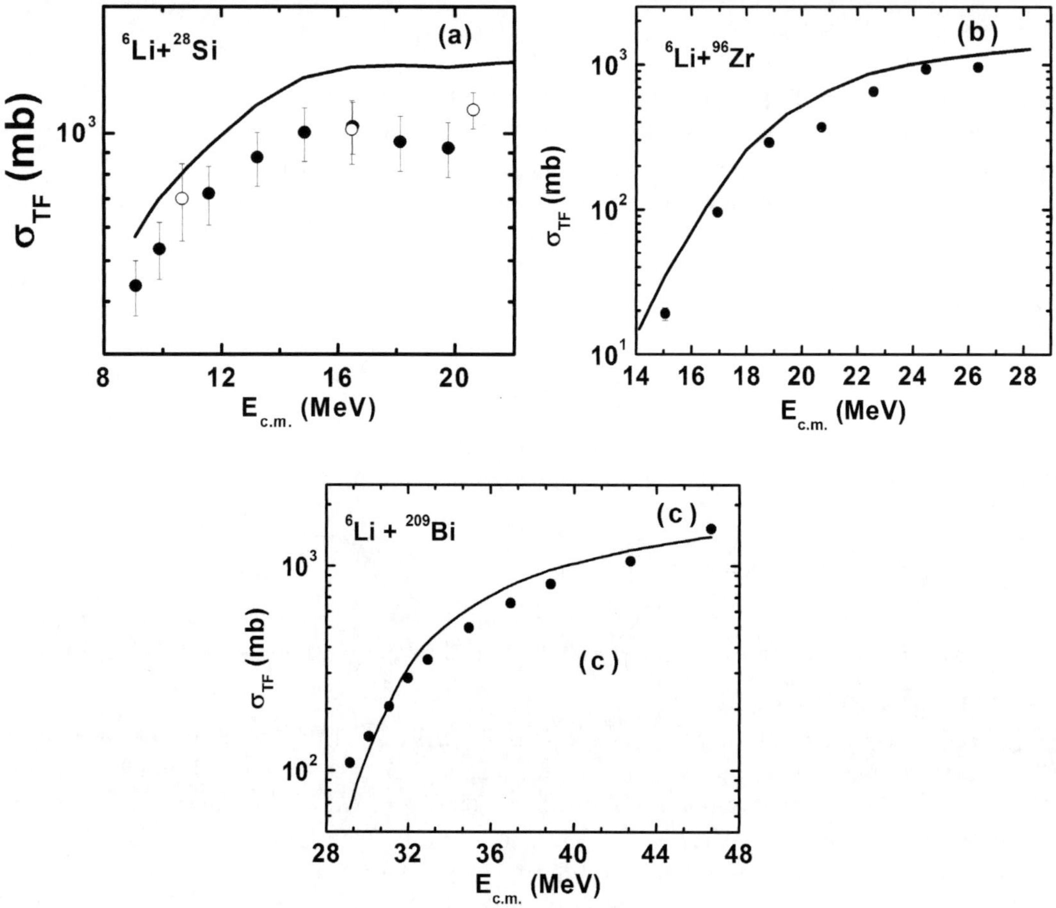


**Figure 1.** CDCC total fusion cross section calculations (solid-lines) induced by $^{6}$Li with targets $^{28}$Si, $^{96}$Zr and $^{209}$Bi. In the calculations, the complete discretized breakup space of the projectile is used.

$r_{dT} = r_{\alpha T} = 1.0$ fm, diffuseness $a_0 = 0.1$ fm and strength $W_0 = 50$ MeV for all targets. Notice that, with these parameters the potentials are inside the Coulomb barrier for all targets.

Figures 1a-1c show the results of total fusion for the $^{6}$Li projectile with targets $^{28}$Si, $^{96}$Zr and $^{209}$Bi respectively. The solid-lines correspond to the calculations with couplings with the full discretized breakup space. The data shown for $^{28}$Si are those of Ref. [38] (solid dots) and [39] (empty dots), for $^{96}$Zr of Ref. [40] and for $^{209}$Bi of Ref. [41]. Regarding the sensitivity of the TF calculations respect to the potential parameters of the short-range potentials, we noticed that smaller values of the reduced radius parameter do not have an appreciable effect as long as the potential is well inside the Coulomb barrier. As for the strength $W_0$ and diffuseness $a_0$, there is not appreciable effect if the range of the potential is larger than the mean free path of the fragment inside the barrier [16].

It is observed in Fig. 1 that the calculations are above the data for $^{28}$Si nucleus, while for the target $^{96}$Zr, the data [40] include lower limits of incomplete fusion measurements, therefore a larger total fusion data is expected. For the nucleus $^{209}$Bi the calculated TF agrees reasonably well with the data. It is important to mention that, we have used systematic nuclear potentials for the interactions of the fragments $\alpha$ [36, 37] and deuteron [35] with the targets. These global potentials depend on the incident fragment energy and are the result of a large systematic study of elastic scattering data in a large range of target masses. So, for that reason some disagreement with the data may appear. Also, as mentioned above, we assume that breakup of $^6$Li occurs into the $\alpha$ and deuteron channel. Although, this is the main dissociation channel, other important mechanisms [27], like $d \to p + n$, with $n$-transfer to the target, are very important. In order to have a complete CDCC description, these breakup possibilities should be included in conjunction with the one treated here. However, treatment of deuteron breakup, with a subsequent neutron transfer, constitutes a four-body calculation which computationally is very difficult to achieve.

Respect to the effect on total fusion of resonance ($l = 2$, $j^\pi = 3^+, 2^+, 1^+$) and nonresonance discretized breakup states of $^6$Li, use is made of the same procedure as in Refs. [32, 33], where a study of the effect of these resonances on elastic scattering was presented. Two different types of calculations are performed; a) fusion is calculated when resonance states are excluded from the full discretized energy space and, b) only couplings between the resonance sub-space are considered. The results are shown in Fig. 2a-2c for $^{28}$Si, $^{96}$Zr and $^{209}$Bi respectively. The solid-lines represent the calculations of Figs. 1a-1c, with couplings with the full discretized energy space. The dashed-lines correspond to the case when only couplings between resonance states are included, while the dashed-dotted lines when resonance states are omitted from the full discretized space. The calculations of fusion through the elastic-channel, i.e., fusion without breakup of the projectile, correspond to the dotted-lines. These are obtained when only the term $\beta' = \beta = 0$ is considered in Eq. (3) and quantify the effect on fusion from the breakup of $^6$Li. In all cases, the elastic-channel calculations are above those with couplings between the complete breakup space. This means that, when breakup states are considered, fusion is suppressed. Thus, a net repulsive polarization potential emerges from couplings to breakup discrete states, which in turn increases the nominal Coulomb barrier and hence suppresses fusion. Now, when couplings between resonance states are considered (dashed-lines), the calculations are closer but still above those with the full space. Respect to the elastic-channel calculation (dotted-lines), resonance couplings produce repulsive polarizations for all targets and energies but still some repulsive contribution should come from nonresonance states. In fact, for the heavier target $^{209}$Bi, the *resonance* calculations are very similar to those with the full discretized space. Hence, couplings to these states become increasingly important as the target mass increases.

It is also observed in Figs. 2a-2c, that for $^{28}$Si, the calculation with couplings between only nonresonance states (dashed-dotted line), shows suppression respect to the elastic-channel (dotted-lines). That is, a net repulsive polarization potential appears from these couplings at energies above the barrier. For $^{96}$Zr at the lowest energies some enhancement is observed. That is, the polarization potential turns attractive, which means that the nuclear component of this potential dominates over the repulsive. For the target $^{209}$Bi, this effect is more evident. At high energies, the nonresonance calculation is very similar to the elastic-channel. However, as the energy decreases towards the barrier, an increasing enhancement appears respect to the elastic-channel calculation. As a matter of fact, the increasing attractive character of nonresonance couplings is compensated by a stronger repulsive polarization from resonance states as observed in Figs. 2b and 2c.

In order to quantify the effect of resonance and nonresonance breakup states of $^6$Li on fusion, we define the ratio $\Gamma$ by,

$$\Gamma_i = 1 - \frac{\sigma_{TF}}{\sigma_{F_i}}, \tag{6}$$

being $\sigma_{F_i}$, fusion through the elastic channel $i = 1$, from resonance $i = 2$ and nonresonance $i = 3$ states. $\sigma_{TF}$ is total fusion with the full discretized breakup space. Figs. 3a-3c show the results of $\Gamma_i$ in terms of $E_{c.m.}/V_B$. The effect of the breakup of the projectile $\Gamma_1$, is shown by the dotted-lines. The dashed-lines correspond to resonance states $\Gamma_2$, while the dashed-dotted lines from

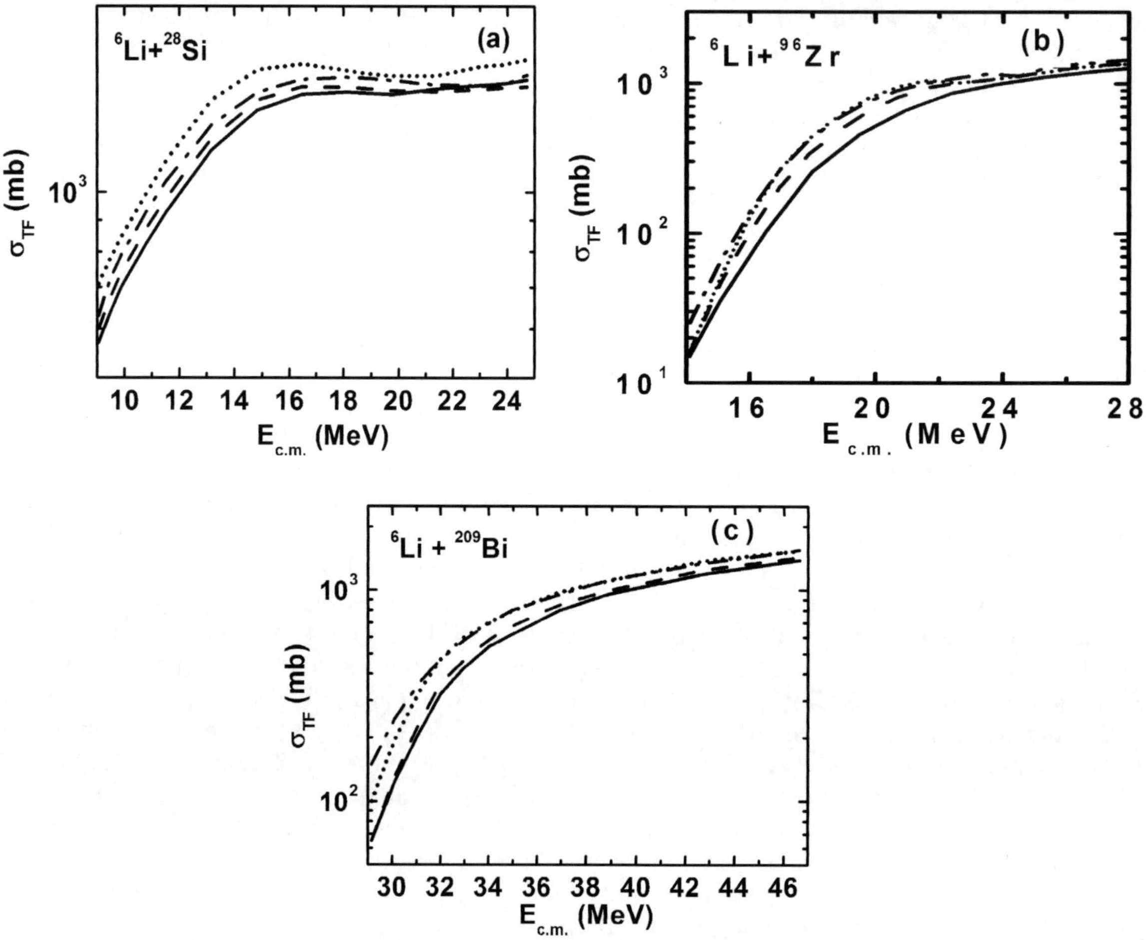


**Figure 2.** Total fusion calculations of $^6$Li with targets $^{28}$Si, $^{96}$Zr and $^{209}$Bi for the elastic channel (dotted-lines), resonance(dashed-lines) and nonresonance (dashed-dotted-lines) couplings. The solid-lines correspond to the calculations with the full breakup space of the projectile

nonresonance states $\Gamma_3$. In these figures, positive values of $\Gamma_i$ mean fusion enhancement respect to the calculation with the full breakup space, while negative values indicate suppression. For all nuclear systems, $\Gamma_1$, $\Gamma_2$ and $\Gamma_3$, increase as the energy decreases towards the barrier $V_B$ to then decrease for lower energies. Also, $\Gamma_2$ has lower values than $\Gamma_1$ and $\Gamma_3$, that is fusion through resonance states is closer to the full calculation. In fact, for the heavier targets $\Gamma_2$ maintains values close to zero and even becomes negative at sub-barrier energies. For $^{96}$Zr, $\Gamma_3$ remains close to $\Gamma_1$, this shows that couplings to these states have a small effect on fusion. However for the heavier target $^{209}$Bi, $\Gamma_3$ becomes even larger than $\Gamma_1$ for energies below the barrier. That is, couplings to nonresonance states strongly enhances fusion. This effect is correlated to the increasing suppression effect produced by resonance states in this energy region.

In order to understand the particular effect on fusion of a given resonance of the projectile, we show in Fig. 4 three types of calculations, namely, fusion is calculated when the states of a single resonance are extracted from the resonance sub-space. The short-dashed-line corresponds

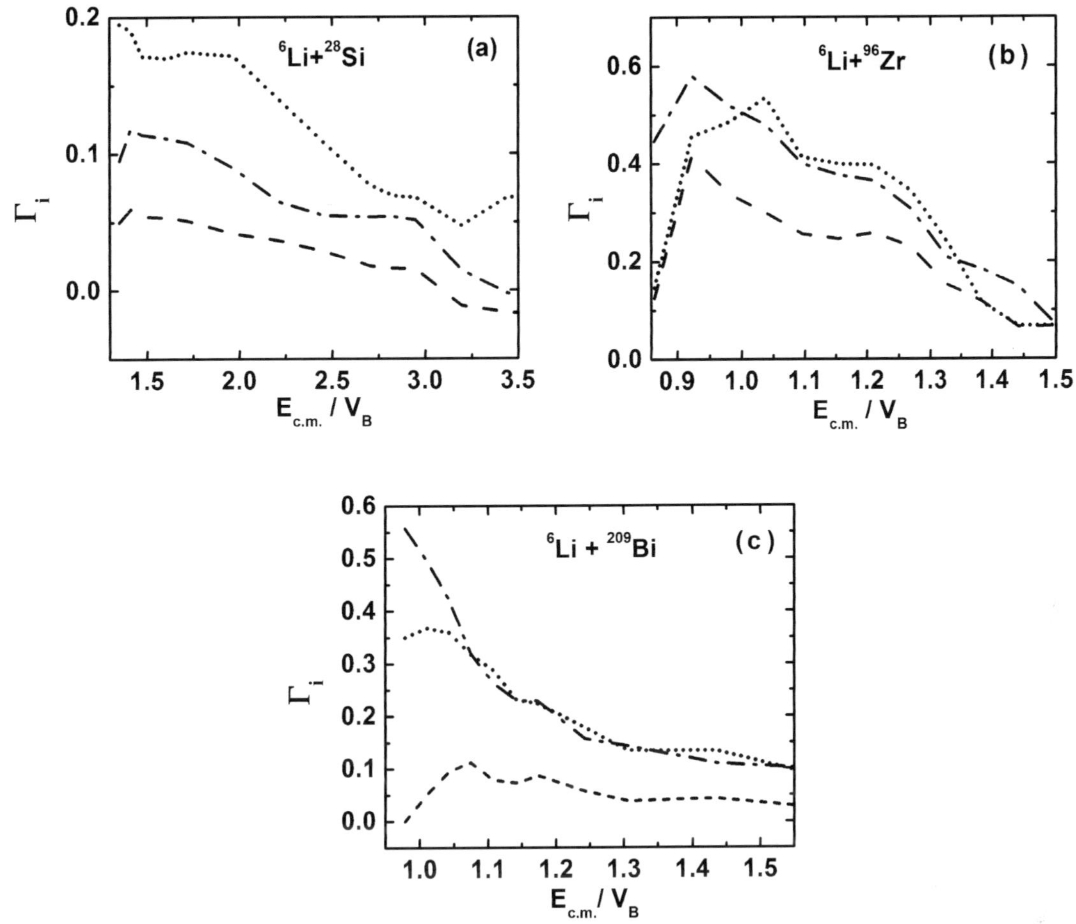


**Figure 3.** Effects on fusion $\Gamma_i$ as defined in Eq. (6), for elastic (dotted-lines), resonance (dashed-lines) and nonresonance (dashed-dotted-lines) sub-spaces.

to the case when the $3^+$-state is extracted from the resonance sub-space. That is, the $2^+$ or $1^+$ are included. The dashed-dotted- and dotted-lines present the results when the $2^+$ and $1^+$ resonances are respectively excluded from the resonance sub-space. The solid-line represents total fusion with the full breakup space (resonance and nonresonance) while, the long-dashed-line with the resonance sub-space. For the $^{28}$Si-target, we observe that for energies above the barrier, the effect of the $3^+$-resonance state is stronger than for the $2^+$ and $1^+$ resonances. That is, the absence of couplings to the $3^+$-state (short-dashed-line) produces a larger enhancement respect to the calculation with the full resonance sub-space (long-dashed-line) and respect to the other two calculations (dashed-dotted and dotted-lines) in which it is included. It is known that for reactions with weakly bound projectiles, total fusion is dominated by complete fusion at high energies, while at energies around and below the barrier, incomplete fusion becomes increasingly important. Contrary to the elastic scattering process [32], the fusion mechanism strongly depends on the incident excited states of the projectile. For example, for nonresonance states prompt breakup takes place, thus these states have a more significant role on NCBU and on ICF processes, particularly at energies around and below the barrier. As for resonance states,

IOP Conf. Series: Journal of Physics: Conf. Series **1078** (2018) 012003 doi:10.1088/1742-6596/1078/1/012003

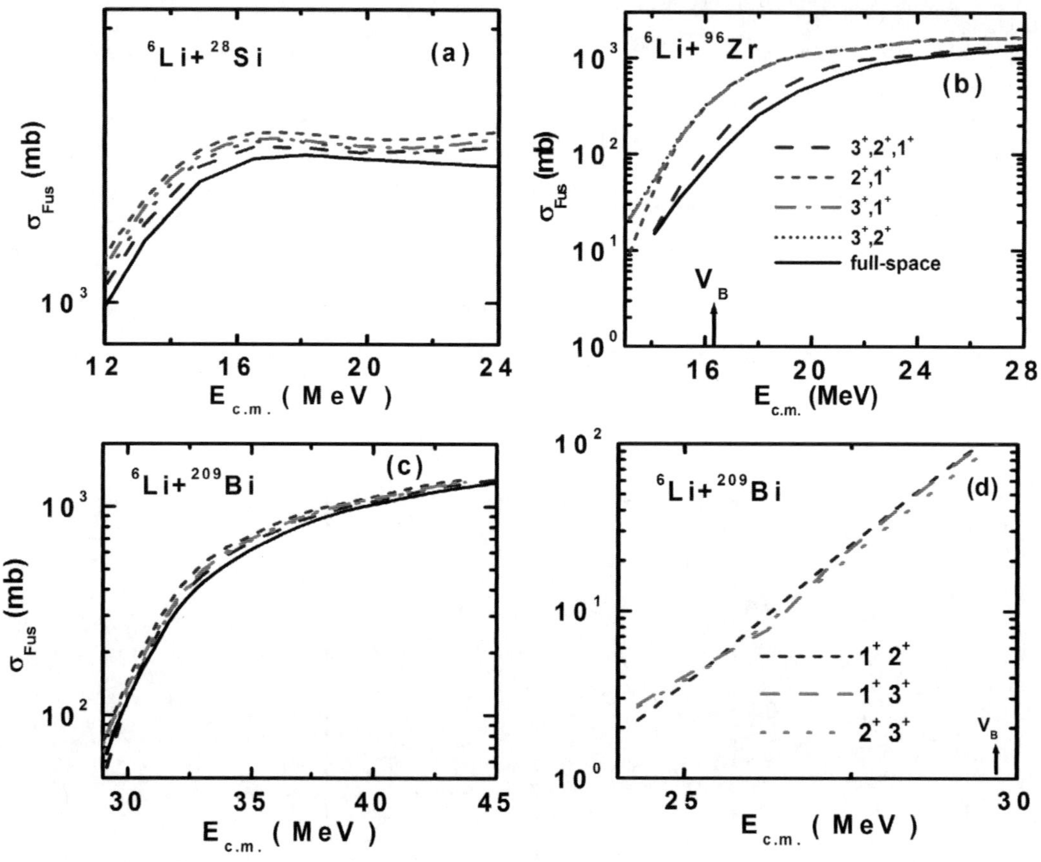


**Figure 4.** Effect on total fusion obtained by the extraction of a single resonance from the resonance breakup space of the projectile.

they have very particular effects on fusion due to the mean-half lives. That is, since the typical collision time is of the order of $10^{-21}$ s, resonance states $2^+$ and $1^+$ with half-lives $3.7 \cdot 10^{-22}$ s and $4.2 \cdot 10^{-22}$ s, can have a more important effect on incomplete fusion than on complete fusion. On the other hand, the $3^+$ resonance, with a longer half-life $262.1 \cdot 10^{-22}$ s, dissociates later and can contribute more to complete fusion.

This is the reason why, as seen in Fig. 4a for the $^{28}$Si-target, when the $3^+$state is considered the results are closer to the TF calculation (solid-line) at energies above the barrier. For the $^{96}$Zr-target as seen in Fig. 4b, the extraction of any single state from the resonance sub-space has similar effects at energies above the barrier, however below the barrier the extraction of the $3^+$-resonance has a more significant effect on fusion. That is, fusion through the $2^+$- and $1^+$-resonances (short-dash-line) approaches the TF values (solid-line) where fusion is dominated by incomplete fusion. For the $^{209}$Bi-target (Fig. 4c) at energies above the barrier, the inclusion of the $3^+$-state gives results closer to TF (solid-line), but at energies well below the barrier, as seen in Fig. 4d, the calculations with the $2^+$- and $1^+$-resonances becomes closer to TF. This findings agrees with those of Ref. [26] for the $^6$Li+ $^{209}$Bi system.

## 3. Summary

CDCC calculations of total fusion cross section have been studied for the weakly bound nucleus $^6$Li with targets $^{28}$Si, $^{96}$Zr and $^{209}$Bi at energies above and below the nominal Coulomb barriers. In the cluster structure scheme of $^6$Li, imaginary short-range potentials have been used to account for absorption of the fragments $\alpha$ and $d$ by the target. When both or one of the fragments are/is inside the range of the absorption potential are/is captured and contribute to fusion. So, complete and incomplete fusion are accounted for in the calculations.

The effects of resonance/nonresonance breakup state couplings of the projectile on fusion have been determined by restricting the calculations to separate resonance and nonresonance breakup sub-spaces. It has been found that resonance states produce a strong effect by suppressing fusion, in fact, this effect is stronger for the heavier targets. At energies above the barrier where, fusion is dominated by the complete fusion process, couplings to the $3^+$-resonance have a significant importance. However, at very low energies where total fusion is dominated by incomplete fusion, the $2^+$ and $1^+$ states have a more important role than the $3^+$-state.

## Acknowledgments

Financial support: A. Gómez from CONACYT, México.

## References

[1] Canto L.F, Gomes P.R.S., Donangelo R., Hussein M.S., Phys. Rep. **424**, 1 (2004)
[2] Liang J.F, Signorini C., Int. J. Mod. Phys. E, **14**, 1121 (2005)
[3] N. Keeley N. , Raabe R., Alamanos N., Sida J.L, Prog. Part. Nucl. Sci. **59**, 579 (2007)
[4] Canto L.F, Gomes P.R.S, Donangelo R., Lubian J., Hussein M.S, Phys. Rep. **596**, 1 (2015)
[5] Gomes P.R.S., et al., Few-body Systems, **57**, 165 (2016)
[6] Dasgupta M., Hinde D.J., Butt R.D., et al., Phys. Rev. Lett. **82**, 1395 (1999)
[7] Signorini C., Liu Z.H., Li Z.C., et al., Eur. Phys. J. A, **5**, 7 (1999)
[8] Gomes P.R.S., Padron I., Crema E., et al., Phys. Lett. **B634**, 356 (2006)
[9] Dasgupta M., Hinde D.J., Hagino K., et al., Phys. Rev. C, **66**, 041602(R) (2002)
[10] Dasgupta M., Gomes P.R.S., Hinde D.J, et al., Phys. Rev. C, **70**, 024606 (2004)
[11] Sakuragi Y., Yahiro M., Kamimura M., Prog. Theor. Phys. Suppl., **89**, 1 (1986)
[12] Sakuragi Y., Yahiro M., Kamimura M., Prog. Theor. Phys. **70**, 1047 (1983)
[13] Austern N., Iseri Y., Kamimura M., et al., Phys. Rep., **154**, 125 (1987)
[14] Hagino K., Vitturi A., Dasso C.H., Lenzi S.M., Phys. Rev. C, **61**, 037602 (2000)
[15] Diaz-Torres A., Thompson I.J., Phys. Rev. C, **65**, 024606 (2002)
[16] Diaz-Torres A., Thompson I.J., Beck C, Phys. Rev. C, **68**, 044607 (2003)
[17] Otomar D.R., Gomes P.R.S., Lubian J., Canto L.F., Hussein M.S., Phys. Rev. C **87**, 014615 (2013)
[18] Beck C., Keeley N., Diaz-Torres A., Phys. Rev. C **75**, 054605 (2007)
[19] Keeley N., Kemper K.W., Rusek K., Phys. Rev. C **65**, 014601 (2001)
[20] Rusek K., Alamanos N., Keeley N., Lapoux V., Pakou A., Phys. Rev. C, **70**, 014603 (2004)
[21] Marta H.D., Canto L.F., Donangelo R., Phys. Rev. C, **89**, 034625 (2014)
[22] Hashimoto S., Ogata K., Chiba S., et al, Prog. of Theor. Phys., **122**, 1291 (2009)
[23] Thompson I.J., Diaz-Torres A., Prog. Theo. Phys. Supp. **154**, 69 (2004)
[24] Parkar V.V., Jha V., Kailas S., Phys. Rev. C, **94**, 024609 (2016)
[25] Gómez Camacho A., Lubian J., Zhang H.Q., et al., Chinese Phys. C, **41**, 124103 (2017)
[26] Diaz-Torres Alexis, Quraishi Daanish, Phy. Rev. C **97**, 024611 (2018)
[27] Luong D.H, Dasgupta M., Hinde D.J, et al., Phys. Rev. C, **88**, 034609 (2013)
[28] Boselli Maddalena, Diaz-Torres Alexis, J. Phys. G, Nucl. Part. Phys., **41**, 094001 (2014)
[29] Boselli M., Diaz-Torres A., Phys. Rev. C, **92**, 044610 (2015)
[30] Diaz-Torres A., Hinde D.J., Tostevin J.A., Dasgupta M, Gasques L.R., Phys. Rev. Lett., **98**, 152701 (2007)
[31] Diaz-Torres A., J. Phys. G, Nucl. Part. Phys. **37**, 075109 (2010)
[32] Gómez Camacho A., Diaz-Torres A., Gomes P.R.S., Lubian J., Phys. Rev. C, **91**, 014607 (2015)
[33] Camacho A.G., Diaz-Torres A., Gomes P.R.S., Lubian J., Phys. Rev. C, **93**, 024604 (2016)
[34] Thompson I.J., Comput. Phys. Rep. **7**, 167 (1988)
[35] An H., Cai C., Phys. Rev. C **73**, 054605 (2006)
[36] Chamon L.C., Pereira D., Hussein M.S., Candido Ribeiro M.A., Galetti D., Phys. Rev. Lett. **79**, 5218 (1997)

[37] Chamon L.C., Carlson B.V., Gasques L.R., et al., Phys. Rev. C **66**, 014610 (2002)
[38] Sinha Mandira, Majumdar H., Basu P., et al., Eur. Phys. J. A. **44**, 403 (2010)
[39] Hugi M., Lang J., Muller R., et al., Nucl. Phys. A **368**, 173 (1981)
[40] Hu S.P., Zhang G.L, Yang J.C., et al., Phys. Rev. C **91**, 044619 (2015)
[41] Dasgupta M., Gomes P.R.S., Hinde D.J., et al., Phys. Rev. C **70**, 024606 (2004)

IOP Conf. Series: Journal of Physics: Conf. Series **1078** (2018) 012004 doi:10.1088/1742-6596/1078/1/012004

# Decoherence Effects in Neutrino Fluxes from the Sun and from Microquasars

**D. R. Bes[1], O. Civitarese[2] and M. E. Mosquera[2,3]**

[1] Department of Physics. Lab.Tandar. CNEA, Argentina
[2] Department of Physics. University of La Plata. c.c.67 (1900) La Plata. Argentina
[3] Faculty of Astronomy and geophysics, University of La Plata, Argentina

E-mail: osvaldo.civitarese@gmail.com

**Abstract.** We study the change which electron-neutrinos may experience during their trip from solar, galactic and extra-galactic sources. The formalism is based on the interaction between neutrinos and the environment. We calculate the time evolution of the density matrix and set limits on the observability of decoherence for neutrino fluxes produced in the sun, in galactic supernovae and in microquasars.

## 1. INTRODUCTION

Neutrino fluxes produced in electroweak decays in the stars are the proper tools to gain some knowledge about astrophysical events and their dynamics. Examples of this type of events are supernovae explosions, highly energetic jets from binary systems, burning of stars, etc. The neutrinos produced in these events can travel from very distant sources to the detectors, nearly at the speed of light with practically no interactions, a fact that allows to re-construct the mechanism of the initial emitting process. However neutrinos of a given flavor are combinations of neutrino mass eigenstates with non-zero masses, as the existence of neutrino oscillations have demonstrated, though their values or hierarchies are still unknown, except for upper limits [1].

From the quantum mechanical point of view the neutrinos are emitted as pure states, in any of the flavors and they could evolve into mixed states due to decoherence, a process which is determined by the interactions with environment [2, 3, 4]. One may then ask the question about the structure of the mixed states which result from the time evolution of the initial states. We may also identify the way in which decoherence would manifest [5, 6]. In the next sections we shall present a simple model and discuss the consequences of it upon the identification of neutrinos from different sources. Then, we shall extent our discussion to some systems of astrophysical interest like microquasars.

## 2. FORMALISM

We shall start with the simplest case by working in the two-flavor basis, where the mass eigenstates $\mid 1 >$ and $\mid 2 >$ have masses $m_1$ and $m_2$ and the Hamiltonian is defined by the matrix

$$H_m = \begin{pmatrix} m_1 & 0 \\ 0 & m_2 \end{pmatrix}. \quad (1)$$

The transformed Hamiltonian in the flavor basis is given by the expression

$$H_f = UH_mU^{-1} = \begin{pmatrix} c^2m_1 + s^2m_2 & -cs(m_2 - m_1) \\ -cs(m_2 - m_1) & s^2m_1 + c^2m_2 \end{pmatrix}, \tag{2}$$

where $c$ and $s$ are a short-hand notation for the cosine and sine of the mixing angle $\theta$. To this Hamiltonian we assign the corresponding form in the spin-matrix representation

$$H_f = \frac{(m_1 + m_2)}{2} n_0 \mathcal{I} - \frac{\epsilon}{2}\sigma_z + \frac{\Delta}{2}\sigma_x, \tag{3}$$

where

$$\begin{aligned} \epsilon &= (c^2 - s^2)(m_2 - m_1) \\ \Delta &= -2cs(m_2 - m_1). \end{aligned}$$

The diagonalization of the eigenvalue problem leads to the density matrix, at time $t = 0$

$$\rho(0) = \frac{\mathcal{I}}{2} + \frac{1}{\delta}\begin{pmatrix} \frac{\epsilon}{2} & -\frac{\Delta}{2} \\ -\frac{\Delta}{2} & -\frac{\epsilon}{2} \end{pmatrix},$$

with $\delta = \pm\sqrt{\epsilon^2 + \Delta^2}$. The time evolution of the density is given by the operator [6, 7]

$$U_n(t) = e^{-iH_nt/^-h} \tag{4}$$

$$= \mathcal{I}\cos\omega_nt - \frac{i}{2\omega_n}(\epsilon_n\sigma_z - \Delta\sigma_x)\sin\omega_nt$$

$$= \begin{pmatrix} \cos\omega_nt - \frac{i}{2\omega_n}\epsilon_n\sin\omega_nt & \frac{i}{2\omega_n}\Delta\sin\omega_nt \\ \frac{i}{2\omega_n}\Delta\sin\omega_nt & \cos\omega_nt + \frac{i}{2\omega_n}\epsilon_n\sin\omega_nt \end{pmatrix}, \tag{5}$$

with (hereafter $\hbar = 1$) $\omega_n = \frac{\sqrt{(\epsilon_n)^2 + \Delta^2}}{2}$ and $\epsilon_n = \epsilon - B_n$, $B_n$ being the environment's energy states. The states of the environment are taken as discrete, time independent, states and denoted by the subindex n. Moreover, they may be taken as continous variables as explained next. The density matrix at the time $t$ is [7]

$$\rho_n(t) = U_n(t)\rho(0)U_n^\dagger(t). \tag{6}$$

The integrated matrix elements, $n$ is taken as a continuous variable such that $\sum_n B_n \to \int dB$, read

$$\rho_{ij}(t) = \int dB \frac{1}{\sqrt{2\pi}s} e^{-\frac{B^2}{2s^2}} \left[\rho_{ij}(t)\right]_n. \tag{7}$$

As a result of the trace operation, $\rho(t)$ represents a mixed state. Diagonalization of this density matrix yields the probabilities $\pi_\pm$ that the neutrinos are found in the pure states $\rho_\pm$

$$\begin{aligned} \rho(t) &= \pi_+\rho_+ + \pi_-\rho_- \\ \pi_+ - \pi_- &= \left(\cos^2 2\theta + \sin^2 2\theta \exp\left[-(s_N\, t)^2\right]\right)^{1/2} \\ \rho_\pm &= \frac{1}{2}\left(\mathcal{I} \pm \frac{\cos 2\theta}{\pi_+ - \pi_-}\sigma_z \right. \\ &\quad \left. \pm \frac{\sin 2\theta}{\pi_+ - \pi_-}\exp\left[-\frac{(s_N\, t)^2}{2}\right](\cos(t)\,\sigma_x - \sin(t)\,\sigma_y)\right). \end{aligned} \tag{8}$$

The probability to detect a neutrino with mixing angle $\gamma$ at time $t$ is given by

$$\pi_\gamma \;=\tfrac{1}{2}\left(1 + \cos(2\gamma)\cos(2\theta) + \sin(2\gamma)\sin(2\theta)\exp\left[-\tfrac{(s_N t)^2}{2}\right]\cos(t)\right). \tag{9}$$

In the asymptotic limit $t \to \infty$, the non-diagonal terms vanish and the system is left in a mass eigenstates with probability $\pi_0 = \frac{1}{2}(1 + \cos 2\theta)$ and $\pi_1 = \pi_{\gamma=\pi/2} = \frac{1}{2}(1 - \cos 2\theta)$. Thus, the mass eigenstates become pointer states. The asymptotic value is independent of the dispersion $s_N$. If $\gamma = \theta$, the initial mixing angle, then

$$\pi_\theta \;= 1 - \tfrac{\sin^2(2\theta)}{2}\left(1 - \exp\left[-\tfrac{(s_N t)^2}{2}\right]\cos(t)\right). \tag{10}$$

Similar results are found if three mass eigenstates are considered [7]. If the environment is oriented along an arbitrary direction $\alpha$ in the $(x,z)$ plane the interaction becomes

$$H^{(coup)} = \frac{1}{2}\sigma_\alpha \otimes \sum_k g_k \sigma_\alpha^{(k)}\;; \qquad \sigma_\alpha = \sigma_z \cos\alpha + \sigma_x \sin\alpha. \tag{11}$$

We rotate the basis states of the environment, so that $\Psi_n^{(\alpha)}$ are eigenstates of the operator $\sum_k g_k \sigma_\alpha^{(k)}$ with the same eigenvalues $\delta_m \beta_n$. The rotation operator is written

$$R_n\left(-\sigma_z + \beta_n \sigma_\alpha\right) R_n^\dagger = \omega_n\, \sigma_z\;; \qquad R(\gamma) = \exp(i\gamma\sigma_y/2), \tag{12}$$

which requires

$$\cos\gamma = \frac{\beta_n \cos\alpha - 1}{\omega_n} \qquad \sin\gamma = \frac{\beta_n \sin\alpha}{\omega_n}\;; \quad \omega_n = (\beta_n^2 - 2\beta_n\cos\alpha + 1)^{1/2}. \tag{13}$$

We replace, as before, the discrete sum by integrals $\zeta_z(\alpha,t), \zeta_x(\alpha,t), \zeta_y(\alpha,t)$, whose explicit expression are found in [7]. The density matrix is written

$$\rho(\alpha,t) \;=\; \frac{1}{2}\left(\mathcal{I} + \sum_{i=z,x,y} \zeta_i(\alpha,t)\,\sigma_i\right) = \pi_+\rho_+ + \pi_-\rho_-,$$

where

$$\begin{aligned} \pi_+ - \pi_- \;&=\; \left(\sum_i \zeta_i^2(\alpha,t)\right)^{1/2} \\ \rho_\pm \;&=\; \frac{1}{2}\mathcal{I} \pm \frac{1}{2(\pi_+ - \pi_-)}\sum_i \zeta_i(\alpha,t)\,\sigma_i. \end{aligned} \tag{14}$$

The probability of detecting a neutrino at time $t$ in the initial state $\rho_\theta$ is

$$\pi_\theta = \frac{1}{2}(1 + \cos 2\theta\,\zeta_z(\alpha,t) + \sin 2\theta\,\zeta_x(\alpha,t)). \tag{15}$$

Thus, while $\rho(0)$ represents a pure state, $\rho(t)$ represents a mixed state. In the next Section we shall briefly discussed two cases of interest. The details are presented in Refs.[7, 8].

## 3. Results and Discussions

Case a) Let us first consider the case of neutrinos produced outside the solar system. The length associated with one oscillation of the density matrix is of the order $\lambda = \nu hc/\delta_m \approx 2.8 \times 10^6 km\ eV/\delta_m$. This estimation may be compared with the length predicted for periodic oscillations of neutrinos $l \approx 1.3 \times 10^6\ km\ eV/\delta_m$, which is of the same order of magnitude than $\lambda$. Assuming an energy of about 20 MeV one gets $l\ \approx 10^{14}\ km$. From these values one gets a solution consisting of $p = 0.77\ \ 1-p = 0.23$ that is about 23 % of the neutrinos arriving to the earth from galactic distances are represented by the particular linear combination $\frac{1}{\sqrt{2}}(\Psi_\mu + \Psi_\tau)$. Since the distance sun-earth is 1.5 $10^8$ km $<< l$, the decoherence mechanism will not affect neutrinos produced within the solar system, since it will require a distance six orders of magnitude larger.

Case b) We shall then consider decoherence effects in fluxes of neutrinos coming from a microquasar [8]. The time evolution of the occupation number of neutrinos is governed by the equation of motion

$$\imath \dot{\rho_f} \quad = \quad \left[ \frac{M^2c^4}{2E\bar{h}} + \sqrt{2}G_F\, \rho,\, \rho_f \right] , \tag{16}$$

where the squared brackets reads for the commutator, the dot represents the time derivative and $M^2c^4$ is the mass-squared matrix in the flavor basis. The quantity $\rho_f$ is the density matrix in the flavor basis, and $\rho$ is a matrix whose diagonal terms are the neutrino number densities, $E$ stands for the neutrino energy and $G_F$ is the Fermi constant.

The mass-squared matrix in the flavor basis can be obtained by a transformation of the mass-squared matrix in the mass basis $(m^2c^4)$ through the unitary mixing matrix $U$. That is:

$$\begin{aligned} M^2c^4 &= Um^2c^4U^\dagger \\ &= \frac{1}{2}\left(m_1^2+m_2^2\right)c^4\begin{pmatrix} 1 & 0 \\ 0 & 1 \end{pmatrix} + \frac{1}{2}\sin 2\theta\left(m_2^2-m_1^2\right)c^4\begin{pmatrix} 0 & 1 \\ 1 & 0 \end{pmatrix} \\ &\quad -\frac{1}{2}\cos 2\theta\left(m_2^2-m_1^2\right)c^4\begin{pmatrix} 1 & 0 \\ 0 & -1 \end{pmatrix} . \end{aligned} \tag{17}$$

As explained previously, we can write the mass matrix and the matrix $\rho_f$ in terms of Pauli matrices. For two-neutrino mass eigenstates they are written

$$\frac{M^2c^4}{2E\bar{h}} \quad = \quad \frac{1}{2}\mathrm{tr}\left(\frac{m^2c^4}{2E\bar{h}}\right) I + \frac{1}{2}w\bar{B}\cdot\bar{\sigma} , \tag{18}$$

$$\rho_f \quad = \quad \frac{1}{2}\mathrm{tr}\left(\rho_f\right) I + \frac{1}{2}\bar{P}_f\cdot\bar{\sigma} , \tag{19}$$

where $\bar{P}_f$ is the polarization vector in the flavor basis, $w = \frac{\delta m^2}{2E\bar{h}}$ and $\delta m^2 = m_2^2c^4 - m_1^2c^4$ is the mass-squared difference between the mass eigenstates, $\bar{B}$ is an unitary vector which fixes the orientation of the background.

$$\bar{B} \quad = \quad \begin{pmatrix} \sin 2\theta \\ 0 \\ -\cos 2\theta \end{pmatrix} . \tag{20}$$

One can performed a rotation of the coordinate system in the flavor space and choose the direction of $\bar{B}$ in the z axis. This direction does not necessary correspond to the external physical magnetic field. The equation of motion is then re-written as

$$\frac{\partial \bar{P}_w}{\partial t} \quad = \quad (w\bar{B} + \mu\bar{P}) \times \bar{P}_w , \tag{21}$$

where $\mu$ stands for the neutrino-neutrino interaction and

$$\bar{P} = \int \bar{P}_w \, dw \ . \tag{22}$$

The order parameter that measures coherence is defined as the ratio between the modulus of the perpendicular polarization vector at time $t$ and time $t = 0$, that is:

$$R_\theta(t) = \frac{|\bar{P}_\perp(t)|}{|\bar{P}_\perp(0)|}, \tag{23}$$

where $\bar{P}_\perp(t) = \bar{P} - (\bar{P} \cdot \bar{B})\,\bar{B}$. The average in angles can be computed as

$$R(t) = \frac{\int R_\theta(t)\, \mathrm{d}\theta}{\int \mathrm{d}\theta}. \tag{24}$$

The time that neutrinos remain inside the jet can be computed as $R_{max}/c$. This time is of the order of $1.4 \times 10^{-3}$ s and in windy microquasar it is $1.7 \times 10^{-4}$ s. Therefore, inside the microquasar's jet, the decoherence has not effect in the neutrino distribution. For the active-sterile neutrino mixing, we found that the length of the vector $\bar{P}$ is reduced to zero for the non-interacting case. Meanwhile, for small values of the interaction the length of the vector $\bar{P}$ is depleted. For the case of the mixing between active neutrinos, if one fixes the neutrino boundaries in the region of the order of $10^8$ cm the time they spend there is of the order of 1.6 $\times 10^{-5}$ seconds, a time shorter than the time needed to evolve into pointer states. However, if the coupling with sterile neutrinos is activated, the onset of decoherence becomes clear from the results. The same behavior is then expected for the outside region. A more precise determination of the effect requires the comparison between the length of the flavor oscillations and the length of decoherence under the microquasar's conditions [8].

## 4. Conclusions

Assuming neutrinos in an initial (pure) flavor state we have obtained the probability of detecting them in a mixed state by taking into account the environment and using a flavor-flavor schematic interactions. Added to the oscillations between flavor-eigenstates we have a rate of decay $\approx \exp(-\mathrm{s}_\mathrm{N}^2 \mathrm{t}^2/2)$, where the parameter $s_N$ measures the width of the residual distribution of eigenvalues in the environment. The amplitude of the oscillations is affected by the exponential decay and it disappears for values of $s_N$ larger than the inverse of the period. We conclude that the probability of detecting a traveling neutrino in a given flavor state is significantly affected by the environment that it may have crossed along its path from the source, an effect which should be considered in addition to flavor oscillations.

## 5. ACKNOWLEDGMENTS

This work has been partially supported by the CONICET (PIP 282) and by the ANPCYT of Argentina.

[1] M. Agostini *et al.*, GERDA collaboration, *Phys. Rev. Lett.* **111**, 122503 (2013); A. Gando *et al.*, KamLAND-Zen collaboration, *Phys. Rev. Lett.* **110**, 062502 (2013); J. B. Albert *et al.*, EXO-200 collaboration, arXiv:1402.6956 (2014); see also Nature **510**,229 (2014).
[2] Decoherence in astro-neutrinos E. Kh. Akhmedov Nucl.Phys.B 538:(1999) 25-51.
[3] Spin baths N.V. Prokofev and P.C.E. Stamp : Theory of the spin bath. Rep. Prog. Phys. 63 (2000) 669-726.
[4] Pointer states W. H. Zurek, Pointer basis of quantum apparatus: Into what mixture does the wave packet collapse?, Phys. Rev. D 24, (1981) 1516-1525.
[5] Environment effects W. H. Zurek, Environment-induced superselection rules, Phys. Rev. D 26, (1982) 18621880.
[6] Quantum decoherence M. Schlosshauer, Decoherence and the quantum to classical transition (Springer)
[7] D. R. Bes and O. Civitarese (to be published)
[8] M. E. Mosquera and O. Civitarese, arXiv 1708.09714 (submitted to JCAP).

IOP Conf. Series: Journal of Physics: Conf. Series **1078** (2018) 012005 doi:10.1088/1742-6596/1078/1/012005

# Electromagnetic and weak decays of baryons in the unquenched quark model

**Roelof Bijker, Gustavo Guerrero-Navarro and Emmanuel Ortiz-Pacheco**

Instituto de Ciencias Nucleares, Universidad Nacional Autónoma de México, A.P. 70-543, 04510 México, D.F., México

E-mail: bijker@nucleares.unam.mx

**Abstract.** In this contribution, we discuss the electromagnetic and weak decays of baryons in the unquenched quark model and show that the observed discrepancies between the experimental data and the predictions of the constituent quark model can be accounted for in large part by the effects of sea quarks. Finally, the obtained results are discussed in terms of flavor-symmetry breaking.

## 1. Introduction

The constituent quark model (CQM) describes the nucleon as a system of three constituent, or valence, quarks. Despite the successes of the CQM (e.g. masses, electromagnetic couplings, and magnetic moments), there is compelling evidence for the presence of sea quarks from other observables such as the observed flavor asymmetry of the proton, the proton spin crisis, and the systematics of strong decays of baryons.

In the CQM, baryons are described in terms of a configuration of three constituent (or valence) quarks neglecting the effects of pair-creation (or continuum couplings). Above threshold these couplings lead to strong decays and below threshold to virtual higher-Fock components (such as $qqq - q\bar{q}$) in the baryon wave function. The effects of these multiquark configurations (or sea quarks) is studied by unquenching the CQM.

In this contribution, we study the importance of sea quarks for the electromagnetic and weak couplings of baryons.

## 2. Unquenched quark model

In the unquenched quark model (UQM), the effects of sea quarks are included via a $^3P_0$ quark-antiquark pair-creation mechanism [1, 2, 3, 4, 5]. The pair-creation mechanism is inserted at the quark level and the one-loop diagrams are calculated by summing over a complete set of intermediate baryon-meson states. As a result, the baryon wave function is given by a superposition of a valence contribution and higher-Fock components consisting of intermediate baryon-meson configurations

$$\mid \psi_A \rangle = \mathcal{N}_A \left\{ \mid A \rangle + \gamma \sum_{BClJ} \int d\vec{K} k^2 dk \mid BC, l, J; \vec{K}, k \rangle \frac{\langle BC, l, J; \vec{K}, k \mid T^\dagger \mid A \rangle}{M_A - E_B(k) - E_C(k)} \right\} . \tag{1}$$

Here $\vec{k}$ and $l$ denote the relative radial momentum and relative orbital angular momentum of the baryon-meson system BC. The operator $T^\dagger$ creates a quark-antiquark pair in the ${}^3P_0$ state with the quantum numbers of the vacuum [4, 5, 6]

$$T^\dagger({}^3P_0) = -3\gamma \int d\vec{p}_4\, d\vec{p}_5\, \delta(\vec{p}_4+\vec{p}_5)\, C_{45}\, F_{45}\, e^{-\alpha_d^2(\vec{p}_4-\vec{p}_5)^2/8} \left[\chi_{45} \times \mathcal{Y}_1(\vec{p}_4-\vec{p}_5)\right]_0^{(0)} b_4^\dagger(\vec{p}_4)\, d_5^\dagger(\vec{p}_5)\,. \tag{2}$$

The quark-antiquark pair is characterized by a color singlet wave function $C_{45}$, a spin triplet wave function $\chi_{45}$ with spin $S=1$ and a solid spherical harmonic $\mathcal{Y}_1(\vec{p}_4-\vec{p}_5)$ that indicates that the quark and antiquark are in a relative $P$ wave. $F_{45}$ denotes the flavor wave function of the created quark-antiquark pair [7, 8]

$$\frac{1}{\sqrt{2+(m_n/m_s)^2}}\left[|u\bar{u}\rangle + |d\bar{d}\rangle + \frac{m_n}{m_s}|s\bar{s}\rangle\right]\,, \tag{3}$$

which, in the limit of equal quark masses, reduces to the usual expression for a flavor singlet. The ratio of nonstrange to strange quark masses is determined from the quark magnetic moments as

$$\frac{m_n}{m_s} = \frac{m_u+m_d}{2m_s} = \frac{\mu_s(\mu_u-2\mu_d)}{2\mu_u\mu_d}\,. \tag{4}$$

For some observables like the magnetic moments, the effects of quark-antiquark pairs to a large extent can be taken into account by introducing effective (or renormalized) values of the model parameters [4]. As a result, the CQM results are not altered much by the introduction of higher Fock components in baryon (and meson) wave functions. There are other observables for which the explicit inclusion of sea quarks is crucial since the quark-antiquark pairs provide non-neglible contributions that cannot be accounted for by renormalization of model parameters. Examples of the latter are the orbital angular momentum in the spin of the proton [4], the flavor asymmetry of the proton [5], strangeness content of nucleon electromagnetic form factors [3, 9], strangeness suppression [10], and self-energy corrections to baryon and meson masses [11, 12]. In the next section, we discuss the importance of higher-Fock components in electromagnetic and weak decays of baryons.

## 3. Results

In this section, we discuss some recent results for electromagnetic and weak couplings. A more detailed account will be given in future publications [13, 14]. In the present calculation, the sum over intermediate states is limited to octet and decuplet baryons in combination with pseudoscalar octet and singlet mesons. The contributions of radially excited baryons and mesons are not taken into account.

### *3.1. Electromagnetic decays*

The experimental data obtained by the CLAS Collaboration for the electromagnetic decays of $\Sigma$ hyperons of the baryon decuplet show a large deviation from the CQM predictions [15, 16]. Table 1 shows that the experimental widths are underpredicted by almost a factor of two. Here we study the effect of sea quarks for the electromagnetic decays of the decuplet baryons.

The radiative width for this process can be expressed in terms of the transition magnetic moment $\mu_{AB}$

$$\Gamma(A \to B\gamma) = \frac{\alpha E_B p_\gamma^3}{2m_A m_N^2}\mu_{AB}^2\,. \tag{5}$$

The results in Table 1 show that for the $\Delta$ resonance the coupling to the pseudoscalar meson cloud (mostly pions [18]) accounts in large part for the observed discrepancy between the quark model value 399 keV and the experimental value $704 \pm 63$ keV. However, for the $\Sigma$ hyperons the calculated width is larger than the CQM value but still rather far from the experimental result.

**Table 1.** Electromagnetic decay widths in keV.

| | CQM | UQM | Exp [17] |
|---|---|---|---|
| $\Gamma(\Delta \to N\gamma)$ | 399 | 606 | $704 \pm 63$ |
| $\Gamma(\Sigma^{*0} \to \Lambda\gamma)$ | 260 | 318 | $451 \pm 77$ |
| $\Gamma(\Sigma^{*+} \to \Sigma^{+}\gamma)$ | 110 | 131 | $254 \pm 59$ |

### *3.2. Weak decays*

Semi-leptonic decay processes of baryons are described by means of the axial couplings. In the case of octet baryons they can be expressed in terms of the couplings $F$ and $D$. In the quark model their values are given by $F = 2/3$ and $D = 1$. In the Cabibbo approach, $F$ and $D$ are determined from the experimental axial couplings, $g_A(n \to p)$ and $g_A(\Sigma^- \to n)$, leading to effective values, $F = 0.465$ and $D = 0.805$. With these values the semileptonic decay processes of baryons are described very well [19].

In Table 2 we show a comparison of the results for the axial couplings in the CQM and the Cabibbo approach with those of the unquenched quark model. The contribution of the pseudoscalar mesons (and especially that of the pions [18]) is responsable for a substantial lowering of the neutron axial coupling from the CQM value thus bringing it in much closer agreement with experiment without the need to introduce effective values of $F$ and $D$. On the other hand, the result for the $\Sigma^-$ hyperon which is described very well in the CQM is hardly changed. This shows that the effective values of $F$ and $D$ used in the Cabibbo approach can be accounted for in large part by the coupling to the pseudoscalar mesons. A similar conclusion was reached in an earlier study of the meson-cloud model [20].

**Table 2.** Axial couplings for $\beta$ decays.

| | $g_A$ | CQM | UQM | Cabibbo | Exp [17] |
|---|---|---|---|---|---|
| $n \to p$ | $F + D$ | 1.67 | 1.35 | 1.27 | $1.2701 \pm 0.0025$ |
| $\Sigma^- \to n$ | $F - D$ | $-0.33$ | $-0.31$ | $-0.34$ | $-0.340 \pm 0.017$ |
| $\Xi^0 \to \Sigma^+$ | $F + D$ | 1.67 | 1.33 | 1.27 | $1.21 \pm 0.05$ |
| $\Lambda^0 \to p$ | $\frac{1}{\sqrt{6}}(3F + D)$ | 1.22 | 0.93 | 0.90 | $0.879 \pm 0.018$ |
| $\Sigma^- \to \Lambda^0$ | $\sqrt{\frac{2}{3}}D$ | 0.82 | 0.71 | 0.66 | $0.60 \pm 0.03$ |
| $\Xi^- \to \Lambda^0$ | $\frac{1}{\sqrt{6}}(3F - D)$ | 0.41 | 0.30 | 0.24 | $0.31 \pm 0.06$ |
| | $F$ | $\frac{2}{3}$ | $\frac{2}{3}$ | 0.465 | |
| | $D$ | 1 | 1 | 0.805 | |

**Table 3.** Flavor-symmetry breaking in transition magnetic moments (top) and axial couplings (bottom).

| | UQM | SU(3) | | Exp | |
|---|---|---|---|---|---|
| $\mu_{\Delta N}$ | 3.29 | 3.29 | $*$ | $3.53 \pm 0.16$ | $\mu_N$ |
| $\mu_{\Sigma^{*+}\Sigma^{+}}$ | $-2.52$ | $-3.07$ | | $-3.50 \pm 0.43$ | $\mu_N$ |
| $\mu_{\Sigma^{*0}\Lambda^{0}}$ | 2.52 | 2.85 | | $3.03 \pm 0.27$ | $\mu_N$ |
| $g_A(n \to p)$ | 1.35 | 1.35 | $*$ | $1.2701 \pm 0.0025$ | |
| $g_A(\Sigma^- \to n)$ | $-0.31$ | $-0.31$ | $*$ | $-0.340 \pm 0.017$ | |
| $g_A(\Xi^0 \to \Sigma^+)$ | 1.33 | 1.35 | | $1.21 \pm 0.05$ | |
| $g_A(\Lambda^0 \to p)$ | 0.93 | 0.97 | | $0.879 \pm 0.018$ | |
| $g_A(\Sigma^- \to \Lambda^0)$ | 0.71 | 0.68 | | $0.60 \pm 0.03$ | |
| $g_A(\Xi^- \to \Lambda^0)$ | 0.30 | 0.29 | | $0.31 \pm 0.06$ | |

## 4. Flavor-symmetry breaking

Finally, we discuss the UQM results for electromagnetic and weak decays in terms of flavor-symmetry breaking effects. In the UQM, the flavor symmetry is broken by the use of the physical masses of baryons and mesons in the energy denominator in Eq. (1) and the ratio of nonstrange to strange quark masses in the created quark-antiquark pair of Eq. (3). The third column of Table 3 shows the results for the $SU(3)$ flavor symmetry limit. In this case, the transition magnetic moments depend on a single coupling which is normalized to $\mu_{\Delta N}$. For the $\beta$ decays there are two independent couplings. Table 3 shows that, whereas for the weak decays the flavor symmetry is broken by less than 5 %, for the transition magnetic moments the breaking is much larger $\sim 15 - 25$ %.

## 5. Summary and conclusions

In this contribution, we studied the importance of higher-Fock components (or sea quarks) in electromagnetic and weak decays of baryons in the framework of the unquenched quark model. It was shown that the observed discrepancies between the experimental data and the predictions of the CQM can be accounted for in large part by the contributions of quark-antiquark pairs in the UQM. Moreover, it was found that the effects of flavor-symmetry breaking in the UQM are much larger for electromagnetic decays than they are for weak decays.

## Acknowledgments

This work was supported in part by grant IN109017 from DGAPA-UNAM, Mexico and grant 251817 from CONACyT, Mexico.

## References

[1] Törnqvist N A 1985 *Acta Phys. Polon. B* **16** 503
[2] Zenczykowski P 1986 *Ann. Phys. (N.Y.)* **169** 453
[3] Geiger P and Isgur N 1997 *Phys. Rev. D* **55** 299
[4] Bijker R and Santopinto E 2009 *Phys. Rev. C* **80** 065210
[5] Santopinto E and Bijker R 2010 *Phys. Rev. C* **82** 062202(R)
[6] Roberts W and Silvestre-Brac B 1992 *Few-Body Systems* **11** 171
[7] Kalashnikova Y S 2005 *Phys. Rev. D* **72** 034010
[8] Ferretti F, Galatà G, Santopinto E and Vassallo A 2012 *Phys. Rev. C* **86** 015204
[9] Bijker R, Ferretti J and Santopinto E 2012 *Phys. Rev. C* **85** 035204.

[10] Santopinto E, Bijker R and García-Tecocoatzi H 2016 *Phys. Lett. B* **759** 214
[11] García-Tecocoatzi H, Bijker R, Ferretti J and Santopinto E 2017 *Eur. Phys. J. A* **53** 115
[12] Ferretti J and Santopinto E 2014 *Phys. Rev. D* **90** 094022
[13] Guerrero-Navarro G and Bijker R, to be published.
[14] Ortiz-Pacheco E and Bijker R, to be published.
[15] Keller D *et al.* 2011 *Phys. Rev. D* **83** 072004
[16] Keller D *et al.* 2012 *Phys. Rev. D* **85** 052004
[17] Patrignani C *et al.* (Particle Data Group) 2016 *Chin. Phys. C* **40** 100001
[18] Bijker R, Guerrero-Navarro G and Ortiz-Pacheco E 2017 PoS (Hadron2017) 053
[19] Cabibbo N, Swallow E C and Winston R 2003 *Annu. Rev. Nucl. Part. Sci.* **53** 39
[20] Holtmann H, Szczurek A and Speth J 1996 *Nucl. Phys. A* **596** 631

IOP Conf. Series: Journal of Physics: Conf. Series **1078** (2018) 012006 doi:10.1088/1742-6596/1078/1/012006

# High precision mass measurements for the astrophysical r-process

**M Brodeur[1], J Kelly[1], B Liu[1,2], and J Maier[1]**

[1] Department of Physics, University of Notre Dame, Notre Dame, IN 46656, USA
[2] Xi'an Jiaotong University, Xi'an, Shaanxi, China

E-mail: `mbrodeur@nd.edu`

**Abstract.** The production of isotopes via the rapid neutron capture process depends sensitively on the atomic mass of the nuclides involved. In light of the recent observation and confirmation of neutron-star mergers as being one of the sites of the r-process, new nuclear data for improved abundance calculations are more urgently needed. To this end, the development of new radioactive ion beam facilities and measurement techniques now extends our reach into unexplored territories. One such facility is the N = 126 beam factory at Argonne National Laboratory. An essential component of this facility will be a multi-reflection time-of-flight mass spectrometer, currently being commissioned at the University of Notre Dame. Mass resolving power of 60,000 using $^{39}K^+$ ions has been achieved. Simulation and continued optimization are underway to improve the efficiency of the device.

## 1. Introduction

Understanding the production of the various elements observed in the Universe has been for a long time a prolific area of research. While the production of the lightest elements via Big Bang, stellar nucleosynthesis, and cosmic ray spallation is mostly understood, the production mechanism of about half of all elements heavier than iron is still under debate. Indeed, beyond the iron peak the various elements are synthesized by three different processes [1]. The slow neutron capture process [2] inside massive stars is responsible for the production of about half of all elements, while the other half comes from the rapid-neutron capture process (rapid since its time scale is on the order of a second), a lesser known process involving very neutron-rich nuclei [3]. The third process, the p-process, accounts for the remaining isotopes.

The r-process is the least known of all nucleosynthesis processes both in terms of its lack of relevant experimental nuclear data and knowledge of its astronomical site. The latter deficiency was mitigated somewhat last year with the first observation of a neutron star merger. It was first observed remarkably in the form of gravitational waves, but its subsequently detected electromagnetic signals were what had significant ramifications for the underlying nuclear physics [4]. The signature transition lines of the lanthanide showed that these heavy elements could be freshly synthesized by the r-process [5]. This important observation essentially confirmed neutron star mergers as being one of the sites of the r-process [6].

The r-process starts from a series of neutron capture from a seed nuclei and proceed until photo-dissociation reaction rates becomes dominant, resulting in increasingly neutron-rich exotic isotopes. An equilibrium between the (n, $\gamma$) and ($\gamma$, n) reactions along the isotopic chain is then achieved with the only exit point being through the $\beta$-decay of the most exotic isotope in the chain. After this point the series of neutron-captures followed by $\beta$-decays continues in higher-$Z$ elements, forming a path far from stability in the chart of the nuclides that is refereed to as the r-process path.

## 2. Importance of mass measurements

Under (n, $\gamma$) and ($\gamma$, n) equilibrium conditions, the relative abundances of adjacent isotopes, given by the Saha equation, depends exponentially on the one-neutron separation energy $S_n(Z, A+1) = m(Z, A) + m_n - m(Z, A+1)$ where $m$ are atomic masses, and $m_n$ is the mass of one neutron. It is therefore expected that, once established, this equilibrium will only be broken when the neutron supply as been exhausted and the nuclei $\beta$-decay to stability, and so the resulting abundances will be particularly sensitive to atomic mass values.

Unfortunately, the r-process runs far from stability, generally beyond the reach of modern radioactive ion beam facilities (RIB) [7]. Hence there is currently very limited data on atomic masses and one has to rely on nuclear mass models to predict mass values along the r-process path. Commonly used mass models in nuclear astrophysics calculations for the r-process include the Finite-Range Droplet Models (FRDM) [8], Hartree-Fock-Bogoliubov (HFB) approach [9], and Duflo-Zuker [10]. While offering mass values consistent with the 2016 Atomic Mass Evaluation (AME2016) [11] and with one-another close to stability, these models diverge far from stability, in the region of the r-process path [12, 13]. These conflicting mass predictions results in variations in the calculated abundance pattern [14, 13].

Over the past few years, increasingly elaborate sensitivity studies on the effect of atomic masses on the abundance pattern produced by the r-process have been performed [15, 16, 17]. This was studied using different mass models, as well as different astrophysical conditions [13]. The sensitivity of other experimental quantities, including neutron capture rates, beta decay rates, and beta-delayed neutron emission probabilities were also undertaken [13]. Among all these various quantities atomic masses were found to have the greatest impact [13]. The latest sensitivity studies were fully self-consistent, where the variation of the uncertainty on one atomic mass was propagated to the other quantities since atomic masses enters in the determination of all of them [17]. As a result of these studies several conclusions can be drawn [13]: first, variations of the atomic mass of a single nuclei by a value as low as 100 keV can have an impact on the abundance pattern in general. Second, the most influential nuclei are located near the closed $N = 82$ and 126 shells as well as the rare-earth region located around $A = 165$. Finally, while the abundance pattern is sensitive on the choice of mass model or astrophysical condition, similar nuclei appear as being the most sensitive.

To minimize the uncertainty in the abundance pattern associated with a single atomic mass, ideally the measurement should be performed down to an uncertainty of less than 100 keV, which for the mass region where the r-process takes place corresponds to a relative uncertainty of less than $\delta m/m \sim 10^{-7}$.

## 3. The accuracy of Penning trap mass spectrometers

The field of precision atomic mass spectrometry can be divided broadly by their uses of indirect and direct techniques [18]. Indirect techniques involve the measurement of reaction or decay $Q$-values, which related to a difference between several masses. On the other hand, in direct techniques only two nuclei are measured: the nuclei of interest and a calibrant (or reference). These direct techniques can be further divided in two groups: time-of-flight spectrometry and frequency measurements [18]. Each mass spectrometry techniques offers its own advantages. For example, single-turn time-of-flight measurements can accomodate the shortest-lived nuclei and nuclei produced at very low yield. But the direct measurements of cyclotron frequency like those done in Penning traps allow for the most precise mass determination, where relative uncertainties on the order of $\delta m/m \sim 10^{-7}$ are routinely achieved [12]. Penning traps are, however, limited to nuclei with half-lives of 100 ms or more and are not accommodating to high levels of contamination of the incoming beam. It should be noted that nuclei with half-lives as short as 8.8 ms have been measured with a Penning trap [19].

It is not sufficient for the atomic mass measurement to be only precise; they also needs to be accurate. Ultimately, one can have confidence in the accuracy of the mass of a nuclei only if it has been measured more than one time, and by different experiments, employing various techniques like those described above, which will have different technique-dependent systematics. Despite being by far the most precise measurement technique available, therefore making its measurements difficult to compare with others, we must still test the consistency of the atomic mass of a given nuclei that has been measured through repeated measurements.

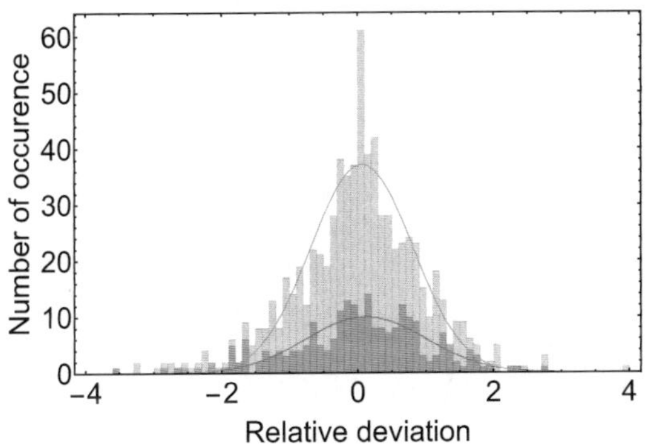


**Figure 1.** Relative deviation between different atomic mass measurements of a given isotope. Yellow: all Penning trap measurements from the AME2016. Dark gray: only the measurements made by different Penning traps. The solid lines represent Gaussian fits of the distributions.

Figure 1 illustrate the consistency of measurement performed using Penning traps. This figure shows the relative deviation of a given mass measurement of a certain isotope versus the most precise Penning trap measurement of that isotope. Essentially, we define the relative deviation $\Delta R_i$ as:

$$\Delta R_i = \frac{m_i(N,Z) - m_0(N,Z)}{\delta m_i(N,Z)}, \tag{1}$$

where $m_i(N,Z)$ is a given mass measurement of a certain isotope, $m_0(N,Z)$ is the most precise measurement of that isotope, and $\delta m_i(N,Z)$ is the uncertainty on the measurement of the isotope of mass $m_i(N,Z)$. Hence $\Delta R_i$ is a measure of how many standard deviations a certain measurement is from the most precise measurement. The values of $m_i(N,Z)$ and $m_0(N,Z)$ where calculated directly from the input values of AME2016 [11]. As shown in Figure 1, most mass measurement are within one standard deviation. If all measurements, including the ones that were measured multiple times by the same group (often times as part of systematic studies), are considered and fitted with a Gaussian distribution, a standard deviation of 0.74 is obtained. The standard deviation is lower than one primarily due to the large number of measurements that were measured by one group multiple times, giving relative deviations close to zero. This experimental setup-related bias in the determination of the consistency of Penning trap measurements can be removed by considering only the isotopes that were measured by more than one experimental setup. The values of $\Delta R_i$ shown in dark color in Figure 1 only consider the masses $m_i(N,Z)$ and $m_0(N,Z)$ measured by different groups. The resulting distribution is more Gaussian and with a standard deviation of 0.88; closer to one. This comparison shows that, in general, measurements performed multiple times by the same experimental setup and different experiment yield consistent values with a nearly statistical spread.

## 4. Penning trap mass spectrometers for the r-process

There are currently seven operational Penning trap mass spectrometer located at RIB facilities in Europe (ISOLTRAP at CERN [20], SHIPTRAP at GSI [21], JYFLTRAP at the University of Jyvaskyla [22], and TRIGATRAP at the TRIGA reactor [23]) and North America (CPT at ANL [24], LEBIT at the NSCL [25], and TITAN at TRIUMF [26]). They are located at facilities with complementary reach in terms of the types of isotope produced and the properties of the beam. Fission so far has been the reaction of choice to produce nuclei of relevance for the r-process since the fission distribution is located around the $N = 82$ shell and the rare-earth region. The JYFLTRAP and CPT systems have been particularly active, with their measurements covering a large portion of the $N = 82$ shell and the rare-earth region [27, 28, 29, 30]. This is because the production method of the radioactive ion beams at their facilities (IGISOL [31] for JYFLTRAP and CARIBU [32] for CPT) is element-independent. While the production mechanism of the radioactive nuclei is different at these two facilities (proton-induced fission on uranium at IGISOL and the spontaneous fission of $^{252}$Cf at CARIBU), for both of them the radioactive nuclei recoil out of the material (target or source) in an ionized form and allowed to thermalize in a helium gas volume. This is in contrast to ISOL facilities [33], where the radioactive nuclei produced in the fission diffuse out of the target only when a high heat is applied. The diffusion process is element specific and typically it is difficult to get refractory element out in sufficient quantities. It should, however, be noted that a small amount of the radioactive refractory $^{54}$V was observed at TITAN [34]. Nevertheless, ISOL facilities have their advantages and sometimes surpass thin-target facilities; their thick targets provide greater opportunity for reactions to occur, and therefore more nuclei of interest can be produced provided they can quickly diffuse out and be ionized. For example, the TITAN Penning trap at ISAC, and ISOLTRAP at ISOLDE, measured very neutron-rich Sr-Rb isotopes [35, 36, 37, 38] of interest for the r-process, as well as neutron-rich Cd [39, 40] and In [41] isotopes more recently.

## 5. Production of critical r-process nuclei at future RIB facilities

Looking in the future, new large user facilities such as FRIB [42] and FAIR [43] will provide more neutron-rich RIBs approaching, and in some cases reaching the r-process path. As an example, Figure 2 shows the mass-measurement reach of the future FRIB facility together with the most sensitive masses for a high-entropy hot wind r-process environment [13]. On this figure, the greater the value of the sensitivity parameter $F_{max}$, the more sensitive is the r-process abundance pattern to that particular atomic mass. Hence, as the figure indicates, a large fraction of critical masses for the r-process would be accessible to the mass spectrometers located at FRIB.

Before the advent of next-generation large-user facilities, one isotopic region remained stubbornly out of reach: $N = 126$ shell closure. This region is of interest for the r-process since it is related to the production of the last abundance peak and, as can be seen in Figure 2, mass uncertainties here have high impact on final abundances. Unfortunately it falls far beyond the two fission distributions of $^{235}$U, and even projectile fragmentation has trouble producing sufficiently exotic nuclei in this region [44]. Fortunately, one can access more neutron-rich in this region by using deep-inelastic reactions [45], where nucleons are exchanged in the near-miss of two stable, similarly heavy neutron-rich species. The KEK Isotope Separation System (KISS) facility at RIKEN [46] is currently employing such a reaction to produce beams near the $N = 126$ shell closure. A second facility using that type of reactions is currently under construction at ANL. Called the $N = 126$ beam factory, it will impinge a $^{136}$Xe beams at 9-10 MeV/u from ATLAS on a $^{198}$Pt target. But here, again, contamination by nearby species remains high, and several types of experiments, such as mass and half-lives measurements, require contaminant-

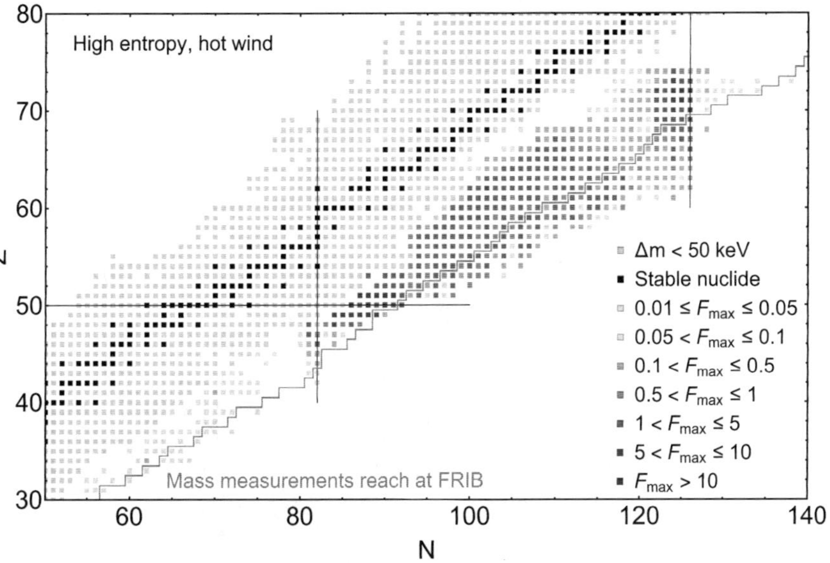


**Figure 2.** In different shades of blue are sensitive masses to the r-process for a high entropy, hot wind scenario. The red line represent the expected reach of mass measurement experiments at FRIB.

free RIBs. Hence, a multi-reflection time-of-flight mass spectrometer (MR-ToF) for isobaric contamination removal as been constructed at the University of Notre Dame [47].

## 6. Off-line commissioning of the ND MR-ToF

The future $N = 126$ beam factory MR-ToF is currently being commissioned off-line at the University of Notre Dame. The dimensions of the device are similar to the ISOLTRAP MR-ToF [48], but with a different assembly design [47]. A schematic of the MR-ToF is shown in Figure 3. It consists of two electrostatic mirrors, each comprising five electrodes. These mirrors are bracketing a long drift tube and two electrodes acting as lenses. The working principle of a MR-ToF is fairly simple: a cocktail bunch containing several different isotopes is captured in the MR-ToF and allowed to bounce back-and-forth between the two mirrors. If the original bunch stays long-enough in the MR-ToF, the various isotopes will start to be resolved according to their mass-to-charge ratios, forming distinct groups. After a certain time the bunch is released from the trap and contaminant ions can be separated by their time-of-flight from the ion of interest. The capture and release of the bunches can be accomplished either by switching the potential of the entrance and exit side mirrors, or by switching the potential on the central drift tube relative to the mirrors [49]. Due to its simplicity, and with the advantage of leaving more electrodes static, the second method was chosen for the Notre Dame MR-ToF.

The off-line test setup, shown on Figure 3, comprises an alkali ion source producing all naturally occurring alkali isotopes from $^{23}$Na to $^{133}$Cs. After the ion source, a Bradbury-Nielsen gate (BNG) is used to chop the continuous beam into short bunches down to 60 ns in length. After, Einzel lenses and steerers are used to focus and align the bunches before injection into the MR-ToF. After the MR-ToF, a second Einzel lens is used to focus the beam on a MCP. Before and after the MR-ToF, Faraday cups and MCPs are placed on actuators such that they can be put in place for diagnosis. The whole system is pumped by four 300 L/s turbo pumps backed

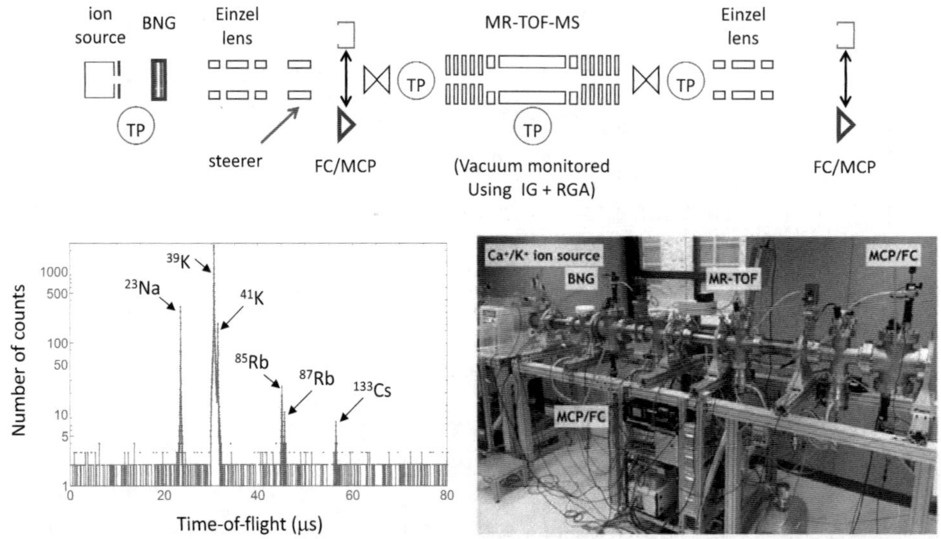


**Figure 3.** Top: Schematic of the beam line for the off-line commissioning of the ND MR-ToF. Bottom left: Time-of-flight spectra of a bunch created by switching the BNG and reaching the last MCP (without being captured). Bottom right: Photo of the beam line at the University of Notre Dame.

by a scroll pump. A vacuum around $2{\times}10^{-10}$ mbar is typically achieved at the location of the MR-ToF.

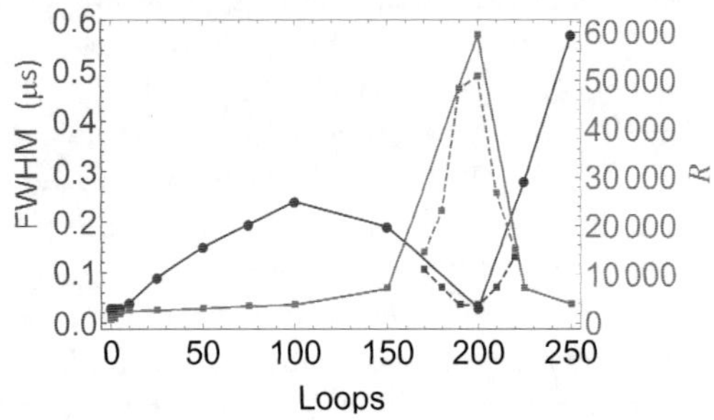


**Figure 4.** Variation of the mass resolving power, $R$, and the full-width half maximum of $^{39}K^+$ bunches as function of the number of round trips.

Figure 4 shows that a mass resolving power reaching 60,000 has been achieved after 200 round trips which is sufficient to resolve isobars. The large increase and peaking of the resolving power at 200 round trip is a consequence of having the time-focus placed at that number of round trips. While the targeted minimum mass resolving power has been reached, the efficiency of the device needs to be improved. To that end, a second steerer has been installed between the BNG and the first Einzel lens. Simulation results show that 100% of the ions from the BNG can be made to reach second MCP through lensing now done with this new steerer, a factor of ten increase compared to the previous configuration. More simulations in conjunction with optimization in the laboratory are currently underway.

## 7. Conclusion

More nuclear physics inputs are needed to better understand the astrophysical r-process (even more critical after the neutron star merger event observation). Atomic masses in particular are one of the most important nuclear data input for r-process abundance calculations. Penning traps are the most precise and accurate mass measurement method to date. Several new masses at Penning traps worldwide have been measured, improving our understanding of the r-process. The last r-process abundance peak around N = 126 is the one with the least experimental information available. The ANL N = 126 beam factory will produce critical nuclei for our understanding of the r-process for the first time and the ND MR-ToF will be used there to remove isobaric contamination from the beam. The MR-ToF is currently being commissioned off-line at Notre Dame and mass resolving powers reaching 60,000 have been observed. Means to improve or understand efficiency losses are being investigated.

## Acknowledgments

The authors would like to acknowledge the assistance of Filip Kondev in providing the AME2016 raw data file of all Penning trap mass measurements. We also acknowledge the support from the University of Notre Dame, and the National Science Foundation under grant numbers PHY-1713857, and PHY-1062819.

## References

[1] Burbidge E M, Burbidge G R, Fowler W A and Hoyle F 1957 *Rev. Mod. Phys.* **29** 547
[2] Käppeler F, Gallino R, Bisterzo S and Aoki W 2011 *Rev. Mod. Phys.* **83** 157
[3] Arnould M, Goriely S and Takahashi K 2007 *Phys. Rep.* **450** 97
[4] The LIGO Scientific Collaboration and the Virgo Collaboration 2017 *Phys. Rev. Lett.* **119** 161101
[5] Chornock R *et al* 2017 *Astrophys. J Lett.* **848** L19
[6] Kasen D, Metzger B, Barnes J, Quataert E and Ramirez-Ruiz E 2017 *Nature* **551** 80
[7] Langanke K and Schatz H 2013 *Phys. Scr.* **T152** 014011
[8] Möller P, Nix J R, Myers W D, Swiatecki W J, 1995 *At. Data Nucl. Data Tables* **59** 185
[9] Goriely S, Chamel N, Pearson J M 2013 *Phys. Rev. C* **88** 024308
[10] Duflo J, Zuker A P, 1995 *Phys. Rev. C* **52** 23
[11] Wang M, Audi G, Kondev F G, Huang W G, Naimi S, and Xu X 2017 *Chin. Phys. C* **41** 030003
[12] Blaum K 2006 *Phys. Rep.* **425** 1
[13] Mumpower M R, Surman R, McLaughlin G C, Aprahamian A 2016 *Prog. in Part. and Nucl. Phys.* **86** 86
[14] Arcones A and Martìnez-Pinedo G 2010 *J. Phys.: Conf. Ser.* **202** 012007
[15] Brett S, Bentley I, Paul N, Surman N and Aprahamian A 2012 *Eur. Phys. J. A* **48** 184
[16] Aprahamian A, Bentley I, Mumpower M, and Surman R 2014 *AIP Advances* **4** 041101
[17] Mumpower M R, Surman R, Fang D-L, Beard M, Möller P, Kawano T, Aprahamian A 2015 *Phys. Rev. C* **92** 035807
[18] Lunney D, Pearson J M, Thibault C 2003 *Rev. Mod. Phys.* **75** 1021
[19] Smith M *et al* 2008 *Phys. Rev. Lett.* **101** 202501
[20] Mukherjee M *et al* 2008 *Eur. Phys. J. A* **35** 1
[21] Block M *et al* 2007 *Eur. Phys. J. D* **45** 39
[22] Szerypo J *et al* 2002 *Nucl. Phys. A* **701** 588
[23] Ketelaer J *et al* 2008 *Nucl. Instrum. Meth. A* **594** 162
[24] Clark J *et al* 2003 *Nucl. Instrum. Meth. B* **204** 487
[25] Ringle R, Schwarz S and Bollen G 2013 *Int. J. of Mass Spectrom.* **349** 87
[26] Dilling J *et al* 2006 *Int. J. of Mass Spectrom.* **251** 198
[27] Van Schelt J *et al* 2012 *Phys. Rev. C* **85** 045805
[28] Van Schelt J *et al* 2013 *Phys. Rev. Lett.* **111** 061102
[29] Kankainen A *et al* 2012 *J. Phys. G: Nucl. Part. Phys.* **39** 093101
[30] Vilen M *et al* 2018, *arXiv:1801.08940 [nucl-ex]*
[31] Jokinen A *et al* 2004 *Nucl. Phys. A* **746** 277c

[32] Savard G *et al* 2008 *Nucl. Instrum. Meth. B* **266** 4086
[33] Blumenfeld Y, Nilsson T and Van Duppen P 2013 *Phys. Scr.* **T152** 014023
[34] Leistenschneider E *et al* 2018 *Phys. Rev. Lett.* **120** 062503
[35] Simon V V *et al* 2012 *Phys. Rev. C* **85** 064308
[36] Manea V *et al* 2013 *Phys. Rev. C* **88** 054322
[37] Klawitter R *et al* 2016 *Phys. Rev. C* **93** 045807
[38] de Roubin A *et al* 2017 *Phys. Rev. C* **96** 014310
[39] Atanasov D *et al* 2015 *Phys. Rev. Lett.* **115** 232501
[40] Lascar D *et al* 2017 *Phys. Rev. C* **96** 044323
[41] C. Babcock *et al* 2018 *Phys. Rev. C* **97** 024312
[42] Hausmann M *et al* 2013 *Nucl. Instrum. Meth. B* **317** 349
[43] Winkler M *et al* 2008 *Nucl. Instrum. Meth. B* **266** 4183
[44] Hirayama Y *et al* 2016 *EPJ Web Conferences* **109** 08001
[45] Watanabe Y X *et al* 2015 *Phys. Rev. Lett.* **115** 172503
[46] Watanabe Y X *et al* 2013 *Nucl. Instrum. Meth. B* **317** 752
[47] Schultz B E, Kelly J M, Nicoloff C, Long J, Ryan S and Brodeur M 2015 *Nucl. Instrum. Meth. B* **376** 251
[48] Wolf R N *et al* 2012 *Nucl. Instrum. Meth. A* **686** 82
[49] Wolf R N *et al* 2012 *Int. J. of Mass Spectrom.* **313** 8

IOP Conf. Series: Journal of Physics: Conf. Series **1078** (2018) 012007 doi:10.1088/1742-6596/1078/1/012007

# Experimental challenges in low-energy nuclear astrophysics

**C.G. Bruno**

School of Physics and Astronomy, The University of Edinburgh, Edinburgh, UK

E-mail: carlo.bruno@ed.ac.uk

**Abstract.** Nuclear astrophysics is an interdisciplinary field at the border between nuclear physics and astrophysics. It aims at answering fundamental questions such as how and where are elements synthesised, how energy is generated in stars and how stars evolve and eventually die. Nuclear astrophysics experiments use Earth-bound facilities to investigate nuclear reactions and nuclear properties of interest in stellar scenarios. This review will focus in particular on thermonuclear reactions occurring in relatively low temperature scenarios (T < 1 GK) such as quiescent burning stars and classical novae. Because of the hindering effect of the Coulomb repulsion between nuclei, nuclear cross sections in these scenarios can be extremely small ($10^{-12}$ barn and even lower) and the signals produced can be challenging to disentangle from the natural background on the Earth's surface.

Moving underground to reduce the background induced by cosmic rays is one possible solution to this problem. For more than 20 years, the LUNA experiment has studied nuclear reactions employing accelerators based deep underground in the Gran Sasso Laboratory in Italy. In this review, recent results obtained at LUNA, as well as future prospects at the new LUNA-MV accelerator, soon to be installed underground will be reviewed.

Another possible approach is to employ indirect methods to probe nuclear properties of astrophysical interest. This review work will mention in particular the novel possibility of carrying out nuclear astrophysics experiments at the newly commissioned CRYRING storage ring in GSI, Germany.

## 1. Introduction

Thermonuclear reactions are responsible both for producing the energy required to keep stars shining and for the synthesis of elements needed for life. Understanding the birth, evolution and death of a star requires precise and accurate knowledge of a great variety of properties of the nuclear species involved in the complex nuclear reaction networks. This review will focus in particular on the nuclear reactions cross sections. Because of the hindering effect of the Coulomb repulsion between charged nuclei, cross sections at the temperatures of interest for low energy nuclear astrophysics (T<1 GK) can be extremely small, of the order of $10^{-12}$ barn or even lower as shown in Fig. 1.

Because the cross section $\sigma(E)$ varies very rapidly in the energy region of interest, it is conventional to express instead the interaction probability using the S-factor S(E), which is implicitly defined from the cross section as

$$\sigma(E) = \frac{1}{E} S(E) \exp^{-2\pi\eta} \qquad (1)$$

where E is the energy (in keV) in the centre-of-mass, and $2\pi\eta = 31.27 Z_1 Z_2 \sqrt{\mu/E}$ is the Sommerfeld parameter, $Z_{1,2}$ are the nuclear charges of the two nuclear involved in the reaction, $\mu$ is the reduced mass and $E$ is the centre-of-mass energy in MeV. The S-factor is useful to study the cross section dependency on nuclear effects, and on nuclear resonances in particular (see Fig. 1).

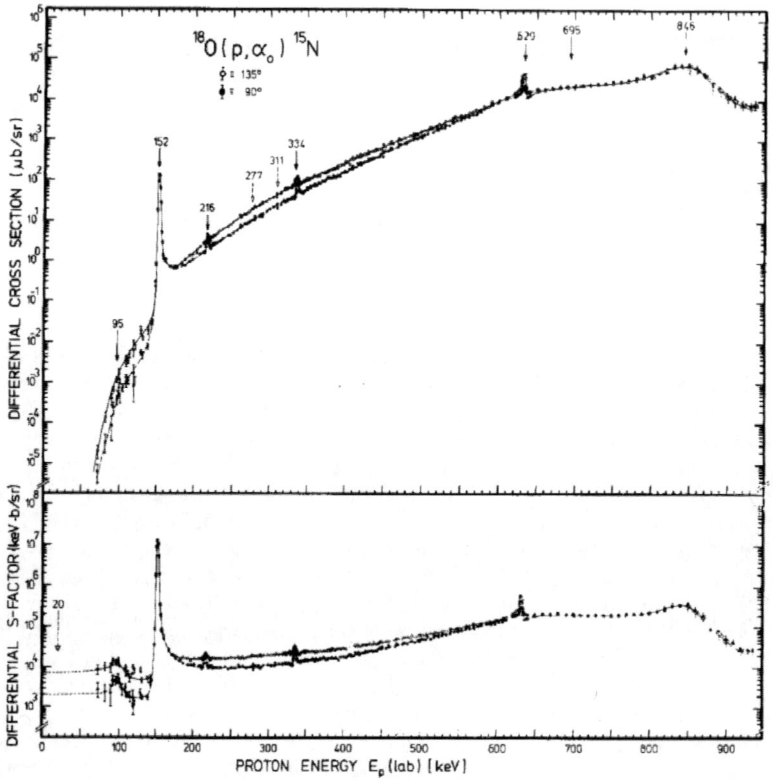


**Figure 1.** Differential cross section (top panel) and differential S-factor (bottom panel) for the $^{18}O(p,\alpha)^{15}N$ reaction. Note the differential cross section is as low as $10^{-5}$ $\mu$barn/sr at the lowest energy point measured. Adapted from reference [1]

Carrying out a direct measurement of a nuclear reaction cross section at energies of astrophysical interest in an Earth-based laboratory can be extremely challenging. As an example, let us consider a proton-induced charge-particle reaction with a cross section of $10^{-12}$ barn. Performing an experiment with a typical beam intensity of 1 $\mu$A and a target thickness of $10^{17}$ atoms/cm$^2$ would result in a yield $Y$ (counts/year)

$$Y = I_{\text{beam}} \times N_{\text{targets}} \times \sigma = 10^{13}\text{pps} \times 10^{17}\text{cm}^{-2} \times 10^{-36}\text{cm}^{-2} \simeq 20\,\text{counts/year} \quad (2)$$

if we consider a detection efficiency of nearly 100%, which could be typical for a charge-particle reaction. If one were to perform a gamma detection experiment instead, typical efficiencies could be around 1% resulting in as little as 0.3 counts/year. Such weak signals require lengthy experimental campaigns and are exceedingly difficult to disentangle from the natural radioactive background. One possible approach to this issue is to carry out the measurement in an underground laboratory, in order to suppress the cosmic component of the natural background. This is the approach taken by the LUNA (Laboratory for Underground Nuclear Astrophysics) collaboration at the underground Laboratori Nazionali del Gran Sasso (LNGS), located in central Italy under the Gran Sasso massif. Another possibility is to employ indirect methods to study the nuclear properties of interest, and this review will focus in particular on the novel possibility of carrying out indirect measurements using the (d,p) technique at the CRYRING storage ring, at GSI Laboratory, Germany.

## 2. Direct underground measurements at LUNA

For more than 20 years, the LUNA collaboration has carried out precise and accurate direct measurements of nuclear reactions underground at LNGS. LUNA started its activity in 1991 using a 50 kV electrostatic accelerator (LUNA-1) [2] which was subsequently replaced in 2002 by a 400 kV machine (LUNA-400kV) [3]. This latter machine is capable of accelerating hydrogen and helium nuclei with typical beam currents of at least 100 $\mu$A, with an beam energy resolution of 0.3 keV and excellent beam stability over time [3]. Fig. 2 shows the location of the LUNA-400kV accelerator, still in use, as well as the former experimental site of the 50 kV machine, and the site for the new LUNA-MV machine (see later).

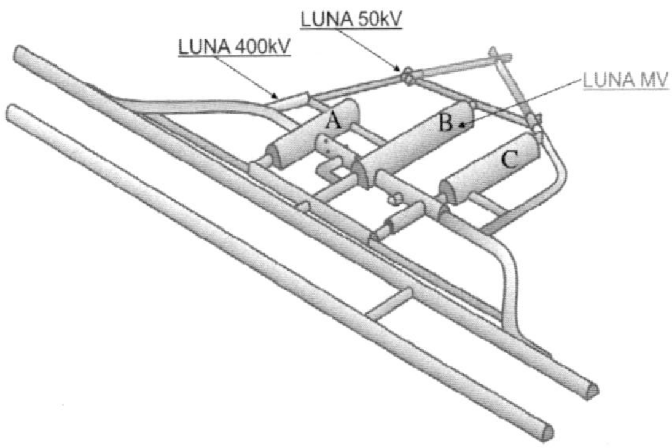


**Figure 2.** A sketch of the underground Laboratori Nazionali del Gran Sasso showing the locations of the LUNA accelerators. See text for details.

Systematic studies were carried out in order to quantitatively assess the background reduction afforded by 1400 meters of rock above the underground laboratory. In particular, Fig. 3 shows the natural background rate measured underground by a HPGe detector surrounded by a lead shield [4] superimposed on the background measured by the same detector above ground. The background suppression achieved is of several orders of magnitude at all gamma energies, greatly enhancing the potential sensitivity of underground gamma spectroscopy experiments.

More recently, similar background comparison tests were performed for silicon detectors used in an alpha spectroscopy experimental campaign [5]. Fig. 4 shows the natural background measured by an array of silicon detectors [6] underground in LNGS and overground in Edinburgh (UK). In both cases, the measurements were carried out with and without a partial lead shield surrounding the setup. Even though the silicon detectors employed in the experimental campaign were too thin (300-700 $\mu$m) for natural background gamma rays to deposit all of their energy, recoil Compton events were shown [6] to contribute significantly to the low-energy background in silicon detectors. At energies lower than 2 MeV, the background reduction afforded by the underground environment is especially significant (over one order of magnitude) and using a lead shield around the setup reduces the background even more. At higher energies, and around 5 MeV in particular, the intrinsic background from radioactive elements in the bulk of the

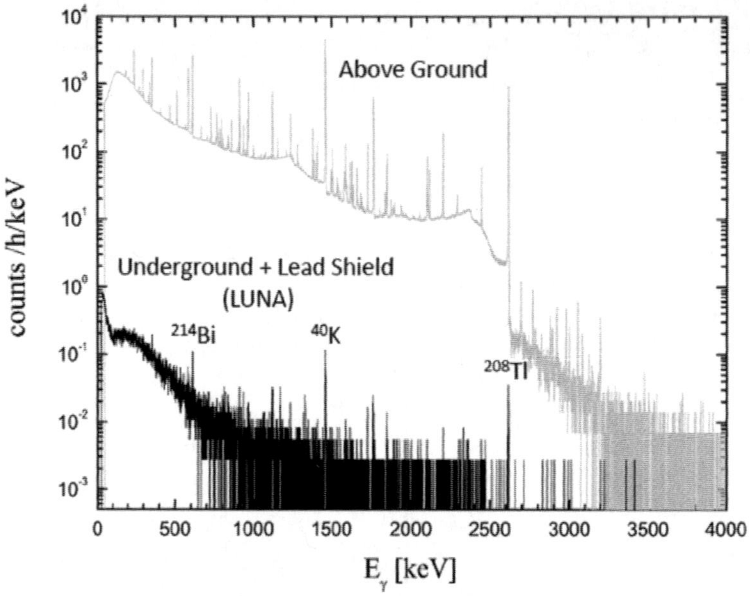


**Figure 3.** Natural background spectrum acquired by a HPGe detector surrounded by a lead shield (black line) and above ground without lead shield (grey line). Adapted from ref. [4]

detector plays a significant role and the advantage of moving underground becomes less obvious. See ref. [6] for more details.

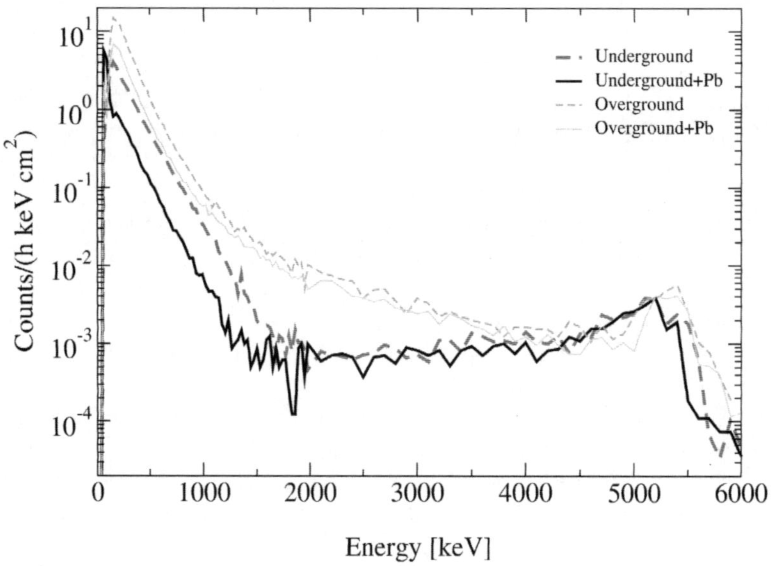


**Figure 4.** Comparison of background spectra taken with and without shielding underground at LUNA and overground in Edinburgh (UK). See text for details. Adapted from ref. [6]

Several studies were carried out at LUNA exploiting the reduction in background afforded

by the underground environment. See ref. [7] for a review. In particular, the original LUNA-1 accelerator was used to carry out a study [8] of the $^{3}$He($^{3}$He,2p)$^{4}$He reaction which was pivotal to understand the (at the time unknown) neutrino oscillations behind the neutrino puzzle. LUNA was able to measure cross sections as low as $20 \times 10^{-15}$ barn and exclude a nuclear solution to the problem [8]. Another highlight of the experimental results obtained at LUNA is the direct measurement of the $^{14}$N(p,$\gamma$)$^{15}$O, the bottleneck reaction of the CNO cycle, at the LUNA-400kV machine [9]. Its revised rate had several significant consequences, in particular in the determination of the age of globular clusters [9].
More recently, the LUNA collaboration has undertaken a systematic campaign aimed at measiring the reactions of the CNO and NeNa cycle. See ref. [10] for a review of the ongoing campaigns and recent results. In particular, a direct measurement $^{22}$Ne(p,$\gamma$)$^{23}$Na reaction [11, 12] revealed several previously unobserved resonances and significantly reduced previous upper limits on unobserved ones, with important implications in a number of stellar sites [13]. Furthermore a recent direct re-measurement of the $^{17}$O(p,$\alpha$)$^{14}$N reaction revealed an enhancement of a factor of two for the strength of a resonance at $E_{CM} = 65$ keV, that dominates the reaction rates at temperatures of astrophysical interest [5]. This measurement had significant implications in a number of astrophysical scenarios [14], and on the identification of stellar grains in particular [15].
A third phase of the LUNA experiment has been approved by the Italian Istituto Nazionale di Fisica Nucleare (INFN) and has already begun construction. A new 3.5MV single-ended accelerator will be installed underground the hall B of LNGS (see Fig. 2). The design goals are high beam intensity and beam stability for proton and helium beams. Among the first reactions to be considered for investigation are the neutron sources $^{13}$C($\alpha$,n)$^{16}$O and $^{22}$Ne($\alpha$,n)$^{25}$Mg.

## 3. Direct and indirect measurements at CRYRING

Direct measurements in direct kinematics, as described in the previous section, are often unsuitable for the investigation of the rates of reactions occurring between short-lived nuclei ($t_{1/2} \ll$ hours), often of great interest for nuclear astrophysics. Measurements in inverse kinematics using in-flight methods of beam production can offer an alternative, but the beam quality is poor and the energies are typically not optimal. A novel approach of using heavy ion beams, decelerated and cooled in-flight in storage rings, was recently pioneered at GSI, Germany [16]. This approach opens up the potential use of high quality beams of previously inaccessible elements/radioactive isotopes [17]. Isotopically pure beams can be used to bombard thin, pure targets (avoiding reactions on contaminants such as carbon [18]), a particularly important feature in studying astrophysical reactions, while maintaining good luminosity due to the re-circulating beam. The pioneering measurement in ref. [16] was carried out at the ESR storage ring in GSI, which is unsuitable for ions circulating at energies significantly lower than 10 MeV/u, typically of interest for nuclear astrophysics. However, the newly installed CRYRING storage ring, currently being commissioned at GSI, would allow ions to circulate even at energies of the order of a few hundred keV/u opening up the possibility of carrying out direct measurements in inverse kinematics using heavy-ion beams in a storage ring.

This section will describe a modular reaction chamber system consisting of two symmetrical interaction chambers for use in nuclear and atomic physics experiments. The two chambers can be mounted either upstream, or downstream, or both upstream and downstream, of a cryogenic jet target. Fig. 5 shows a single chamber mounted upstream. An array of highly segmented Double-sided Silicon Strip Detectors (DSSDs) will be housed in two interaction sub-chambers (Fig. 6) to maximise the angular coverage at very forward (or very backward) angles in the laboratory. The chamber mounted downstream can be used for direct measurements (*e.g.* p,$\alpha$) where reaction products are emitted at forward laboratory angles. The upstream chamber will be used instead for transfer reaction measurements, such as (d,p). In this case, reaction products

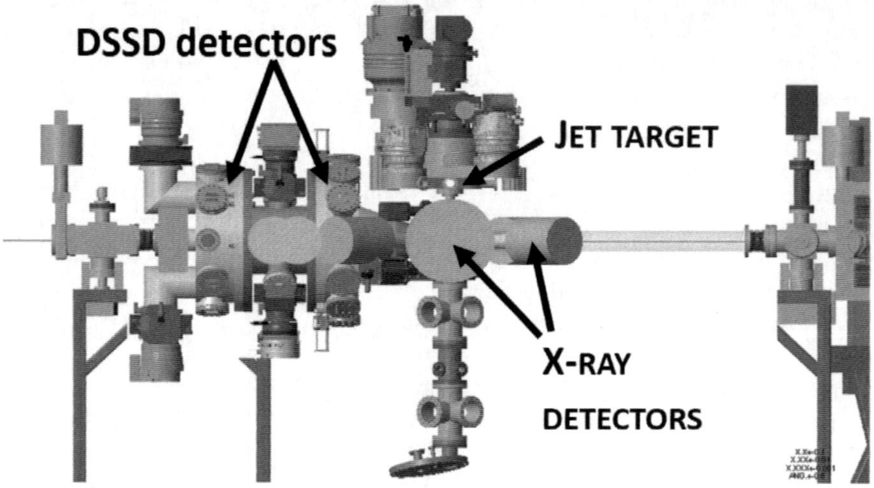


**Figure 5.** A drawing of the nuclear reaction chamber system that will be installed at the CRYRING. Only one chamber is shown, the other chamber could be mounted on the other side of the gas jet target. Note the interaction chambers housing the Double-Sided Silicon Strip Detectors, and the HPGe X-ray detectors mounted around the cryogenic jet target.

**Figure 6.** A drawing of the detector chamber furthest from the jet target. Detectors are placed at approximately ±5 mm from the beam axis to detect recoils or ejecta emitted at very forward/backwards laboratory angles. The beam passes in the central hole between the detectors, which can be moved outwards to increase the beam clearance.

are emitted at forward angles in the centre-of-mass frame, corresponding to backward angles in the laboratory frame. The (d,p) transfer technique is particularly useful in measuring the

energies, the spectroscopic factors $C^2S$ [18], and the spin-parities of low angular momentum transfer states (*i.e.* $l$=0,1) which usually play a key role in astrophysical scenarios.
In particular, a first proposal to measure the $^{30}$P(d,p)$^{31}$P reaction using this chamber was submitted, and accepted, at the 2017 GSI G-PAC. This indirect study aims at investigating $^{31}$P, the mirror partner of the compound nucleus produced by $^{30}$P(p,$\gamma$)$^{31}$S reaction. The $^{30}$P(p,$\gamma$)$^{31}$S reaction plays a key role in the synthesis of all elements heavier than Ca in classical novae (ref. [19] and Fig. 7) and is critical to understand $^{30}$Si/$^{28}$Si isotopic ratios observed in pre-solar grains thought to originate from ONe novae [20]. Pre-solar grains are tiny fragments of material, ejected from novae pre-dating the formation of the Solar system, and brought to Earth by meteorites. An improved knowledge of the rate of the $^{30}$P(p,$\gamma$)$^{31}$S reaction could allow us to unambiguously identify the type of nova from which a given grain originates from, greatly enhancing the current experimental constraints on novae models.

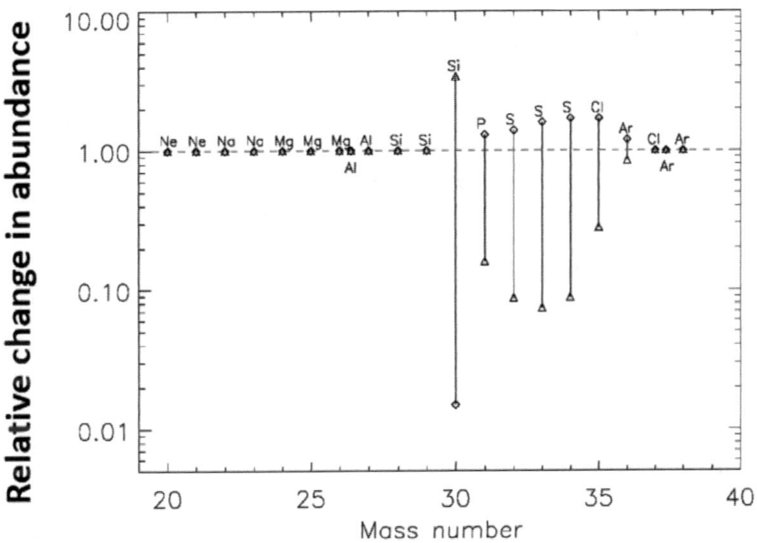


**Figure 7.** Uncertainties in a nova model prediction of the isotopic abundance composition of novae ejecta arising from uncertainties in the $^{30}$P(p,$\gamma$)$^{31}$S reaction rate. Adapted from ref. [19].

## 4. Conclusions
Low-energy nuclear astrophysics aims at measuring the cross section of nuclear reactions taking place in a variety of low-temperatures scenarios, such as quiescent burning and classical novae. These cross sections measurements are extremely challenging measurements because of the hindering effect of the Coulomb barrier. One possible approach is to carry out the measurements underground, in order to suppress the natural background and enhance the signal to noise ratio. The LUNA collaboration has followed this approach to produce high-precision measurements for the last 20 years. The LUNA-MV machine, soon to be installed underground at the Laboratori Nazionali del Gran Sasso, will open up new and exciting prospects for underground nuclear physics. Another possible approach is to carry out the measurements inside a storage ring, employing radioactive beams produced via in-flight techniques. These beams can be used for both direct and indirect measurements. This novel technique will be employed for a variety of nuclear astrophysical studies at the CRYRING storage ring in GSI, Germany.

## 5. Acknowledgements
This work is supported by the UK Science and Technology Facilities Council (STFC).
The author would like to gratefully acknowledge the advice and help of the members of the LUNA collaboration, as well as the ISOL-SRS, CRYRING, and NucAR collaborations.

## References
[1] H.-M. Lorenz-Wirzba *et al.*, Nucl. Phys. A **313** (1979) 346-362
[2] U. Greife *et al.*, Nucl. Inst. Meth. A **350**, 327 (1994)
[3] A. Formicola *et al.*, Nucl. Inst. Meth. A **507**, 609 (2003)
[4] H. Costantini *et al.*, Rep. Prog. Phys **72**, 086301 (2009)
[5] C.G. Bruno *et al.*, Phys. Rev. Lett. **117** (2016) 142502
[6] C.G. Bruno *et al.*, Eur. Phys. J. A **15** (2015) 94
[7] C. Broggini *et al.*, Annu. Rev. Nucl. Part. Sci. **60** (2010) 53
[8] R. Bonetti *et al.*, Phys. Rev. Lett. **82** (1999) 26
[9] G. Imbriani *et al.*, Astron. Astrophys. **420** (2003) 625
[10] A. Boeltzig *et al.*, Eur. Phys. J. A **52** (2016) 75
[11] F. Cavanna *et al.*, Phys. Rev. Lett. **115** (2015) 252501
[12] R. Depalo *et al.*, Phys. Rev. C **94** 055804 (2016)
[13] A. Slemer *et al.*, MNRAS **465** 4817-4837 (2017)
[14] O. Straniero *et al.*, Astron. Astrophys. **598** A128 (2017)
[15] M. Lugaro *et al.*, Nature Astronomy **1** 0027 (2017)
[16] B. Mei *et al.*, Phys. Rev. C **92** 035803 (2015)
[17] M. Lestinksy *et al.*, Eur. Phys. J Special Topics **225** 797 (2016)
[18] V. Margerin *et al.*, Phys. Rev. Lett. **115** 062701 (2015)
[19] C. Iliadis *et al.*, Astrophy. J. Supp. Series **142** (2002)
[20] J. José *et al.*, Astrophys. J **612** (2004) 414

IOP Conf. Series: Journal of Physics: Conf. Series **1078** (2018) 012008 doi:10.1088/1742-6596/1078/1/012008

# Experimental challenges in the measurement of double charge exchange reactions within the NUMEN project

**D. Carbone[1], F. Cappuzzello[1,2], C. Agodi[1], M. Cavallaro[1], L. Acosta[3], D. Bonanno[4], D. Bongiovanni[1], I. Boztosun[5], S. Calabrese[1,2], D. Calvo[6], E.R. Chávez Lomelí[3], F. Delaunay[6,7], N. Deshmukh[1], P. Finocchiaro[1], M. Fisichella[6], A. Foti[4], G. Gallo[1,2], A. Hacisalihoglu[1,8], F. Iazzi[6,9], R. Introzzi[6,9], G. Lanzalone[1,10], R. Linares[11], F. Longhitano[4], D. Lo Presti[2,4], N. Medina[12], A. Muoio[1], J.R.B. Oliveira[12], A. Pakou[13], L. Pandola[1], F. Pinna[6,9], S. Reito[4], G. Russo[2,4], G. Santagati[1], O. Sgouros[1,13], S.O. Solakcı[5], V. Soukeras[1,13], G. Souliotis[14], A. Spatafora[1,2], D. Torresi[1], S. Tudisco[1], A. Yildirim[5], V.A.B. Zagatto[11] for the NUMEN collaboration**

[1] INFN-Laboratori Nazionali del Sud, Catania, Italy
[2] Università degli Studi di Catania, Catania, Italy
[3] Universidad Nacional Autònoma de México, Ciudad de México, Mexico
[4] INFN-Sezione di Catania, Catania, Italy
[5] Akdeniz University, Antalya, Turkey
[6] INFN-Sezione di Torino, Turin, Italy
[7] LPC Caen, Normandie Université, ENSICAEN, UNICAEN, CNRS/IN2P3, Caen, France
[8] Istitute of Natural Science, Karadeniz Teknik Universitesi, Trabzon, Turkey
[9] DISAT, Politecnico di Torino, Turin, Italy
[10] Università di Enna "Kore", Enna, Italy
[11] Universidade Federal Fluminense, Niteroi, Brazil
[12] Universidade de Sao Paulo, Sao Paulo, Brazil
[13] University of Ioannina, Ioannina, Greece
[14] Laboratory of Physical Chemistry, Department of Chemistry, National and Kapodistrian University of Athens, Athens, Greece

carboned@lns.infn.it

**Abstract.** The NUMEN project proposes to measure the absolute cross sections of heavy-ion induced Double Charge Exchange (DCE) reactions with the final goal to get information on the nuclear matrix elements involved in the neutrinoless double beta (0νββ) decay. The knowledge of the nuclear matrix elements is crucial to infer the neutrino average masses from the possible measurement of the half-life of 0νββ decay and also to compare experiments on different isotopes. DCE reactions and 0νββ decay present some similarities, the initial and final-state wave functions are the same and the transition operators are similar. Many challenges have to be faced for the experimental measurements of DCE reactions induced by heavy ions, since they are characterized by very low cross sections.

## 1. Introduction

The physics of neutrinoless double beta (0νββ) decay has fundamental implications since its observation would signal that the total lepton number is not conserved and also it would establish the Majorana nature of neutrino. Therefore, the search of this very rare process is the topic of many research facilities all over the world. The 0νββ decay basically involves nuclei, thus its analysis necessarily implies nuclear structure items. Indeed, the 0νββ decay rate can be expressed as a product of three independent factors: the phase-space factor, the Nuclear Matrix Element (NME) and a term which - in the standard scenario of decay mediated by the exchange of Majorana neutrinos – contains the effective neutrino masses. The precise knowledge of NMEs is thus mandatory to extract information on the neutrino masses, when the decay rate will be possibly measured. It is also a critical ingredient to convert the actual limits on the half-life established by the experiments with different isotopes into limits on the neutrino mass.

Presently, the evaluation of the NMEs is based mainly on state-of-the-art calculations with different approaches (QRPA, shell-model, IBM etc.) [1-4]. However, significant differences in the obtained values are present due to ambiguities in the models and the lack of strong constraints. Moreover, possible common approximations could correspond to systematic uncertainties.

In order to give experimentally driven information on the NMEs, the NUMEN [5-7] and the NURE [8-9] projects started an experimental campaign on heavy-ion induced Double Charge Exchange (DCE) reactions. Even if DCE reactions and 0νββ decay are mediated by different interactions, there is a link between them based on some important similarities: i) the initial and final state wave functions in the two processes are the same, ii) the transition operators are similar, in both cases Fermi, Gamow-Teller and rank-two tensor components are present, iii) a large linear momentum (~100 MeV/c) is available in the virtual intermediate channel, iv) the two processes are non-local and are characterized by two vertices localized in a pair of valence nucleons, v) they take place in the same nuclear medium, vi) a relevant off-shell propagation through virtual intermediate channels is present.

The advantage is that DCE reactions can be induced in laboratory, but a simple relation between DCE cross sections and ββ-decay half-lives is not trivial and needs to be explored.

The experimental activity of the projects is mainly performed at the INFN-Laboratori Nazionali del Sud in Catania using the high resolution Superconducting Cyclotron (CS) beams and the MAGNEX large acceptance magnetic spectrometer, which is characterized by high resolution in energy, mass and angle [10-12]. Indeed, the high-order solution of the equation of motion implemented in MAGNEX guarantees the above mentioned performances and its relevance in the research of heavy-ion physics [13-17], some of them obtained from its coupling to the EDEN neutron detector array [18][19].

We already established the feasibility of such experiments by studying the $^{40}$Ca($^{18}$O,$^{18}$Ne)$^{40}$Ar DCE reaction at 15 AMeV with the aim to measure the absolute cross section at zero degree [20]. This pilot experiment demonstrated that high resolution and statistically significant experimental data can be measured for DCE processes and that precious information towards NME determination could be at our reach [20].

Recently, we focused the experimental activity on DCE reactions involving the nuclei of interest for 0νββ decay. In particular, the ($^{20}$Ne,$^{20}$O) DCE reactions at 15 AMeV on $^{116}$Cd, $^{76}$Ge and $^{130}$Te, were measured for the first time. Some details about the experimental challenges and data reduction procedure of these measurements are discussed in this paper.

## 2. Towards the experiments of the NUMEN and NURE projects

The aim of the NUMEN project is to measure the absolute cross section for DCE reactions on target nuclei candidates for the $0\nu\beta\beta$ decay and find a connection between the NMEs of the two processes. With respect to the pilot experiment, performed on $^{40}$Ca target, DCE reactions on such target nuclei are characterized by some additional experimental difficulties. The challenge is to measure a rare nuclear transition under a very high rate of heavy ions produced by the beam-target interaction. In the exploration of nuclei of interests for $0\nu\beta\beta$ we consider that:

a) The $Q$-value for DCE reactions on such nuclei is typically more negative than in the case of $^{40}$Ca explored in ref. [20]. This could strongly reduce the cross section.

b) The ($^{18}$O,$^{18}$Ne), which emulate the $\beta^+\beta^+$ decays, is particularly advantageous, due to the large value of the B(GT) strengths. To explore reactions of the $\beta^-\beta^-$ kind, a possible candidate is, for example, the ($^{18}$Ne,$^{18}$O), which requires a radioactive beam that cannot be available with comparable intensity. The proposed ($^{20}$Ne,$^{20}$O) has smaller B(GT), so a sensible reduction of the yield could be expected;

d) The study of some nuclei requires gas or implanted targets, e.g. $^{136}$Xe or $^{130}$Xe, which are normally much thinner than solid state ones, with a consequent reduction of the collected yield;
e) In some cases the achievable energy resolution is not enough to separate the ground from the excited states in the final nucleus. Thus, the coincident detection of γ-rays from the de-excitation of the populated states is mandatory, but at the price of the collected yield.
As a consequence, the present limits of beam power (~100 W) for the CS accelerator and acceptable rate for the MAGNEX focal plane detector (few kHz) allow us to concentrate on only few cases, which are planned in the NURE project [8-9] (e.g. $^{116}$Cd, $^{130}$Te, $^{76}$Ge).
In order to start a systematic exploration of all the nuclei of interest for $0\nu\beta\beta$ decay, we need to work at about two orders of magnitude more luminosity than the present and thus an upgraded set-up is necessary. The plan is to develop a substantial change in the technologies implemented in the beam extraction [21], in the control of the beam induced radioactivity, in the detection system of MAGNEX [22-26] and in the power dissipation of the thin targets [27]. In addition, in order to explore a wider range of incident energy, an increase of the maximum accepted magnetic rigidity is foreseen. This will be done preserving the geometry and field uniformity of the magnetic field [28-31], in order to keep the high-precision of the present trajectory reconstruction.
Finally, the development of a specific theory program to allow an accurate extraction of nuclear structure information from the measured cross sections is mandatory and is part of the NUMEN project.

## 3. First experimental measurements on the ($^{20}$Ne,$^{20}$O) reaction

We performed first experimental investigations of the ($^{20}$Ne,$^{20}$O) DCE reaction on $^{116}$Cd,$^{76}$Ge and $^{130}$Te targets, which are nuclei candidates for the 0νββ decay. These are the first measurements of such a reaction, there are no data available in literature. A $^{20}$Ne$^{10+}$ beam was accelerated at 15 AMeV by the CS of INFN-LNS. The thicknesses of the targets were carefully chosen in order to obtain an energy resolution which allows to distinguish the transition to the residual nucleus ground state from its first excited state. A $^{116}$Cd rolled target of 1370 μg/cm$^2$ thickness and $^{76}$Ge (386 μg/cm$^2$ thickness) and $^{130}$Te (247 μg/cm$^2$ thickness) both evaporated on a C backing of ~ 50 μg/cm$^2$ were used. The thickness of $^{116}$Cd is much higher than that of $^{76}$Ge and $^{130}$Te, because the first excited state in the corresponding residual nucleus of $^{116}$Sn is at 1.293 MeV, to be compared to 0.559 MeV in $^{76}$Se and 0.536 MeV in $^{130}$Xe. The MAGNEX spectrometer was placed at forward angles including zero degree in the full acceptance mode (~50 msr).
The magnetic fields were set in order to transport the $^{20}$Ne$^{10+}$ ions towards the faraday cup position at the focal plane. However, when the beam passes through the targets a charge state distribution is originated. The maximum amount corresponds to the fully stripped $^{20}$Ne$^{10+}$ (~ 99%) but a sizeable amount of beam in the 9$^+$ and 8$^+$ charge states is also produced. These lower charge state components have a magnetic rigidity similar to that of the ejectiles of interest: $^{20}$F$^{9+}$ for the Single Charge Exchange (SCE) and $^{20}$O$^{8+}$ for DCE. Consequently, they enter in the FPD acceptance causing a limitation in beam intensity tolerable by the detector. In order to minimize the amount of $^{20}$Ne$^{9+}$ and $^{20}$Ne$^{8+}$ beams, a second target was placed downstream of the primary one to be used as a post-stripper material. Different materials where tested and the final choice was a thick C foil of ~ 800 μg/cm$^2$. With this configuration the charge state distribution is ~ 99.1 % of 10$^+$, ~ 9.0 · 10$^{-3}$ % of 9$^+$ and ~ 2.0 · 10$^{-5}$ % of 8$^+$ [32]. This solution allowed only partially to reduce the background and thus a system of shields before the FPD entrance was also equipped to stop such ejectiles.

## 4. Data reduction

The data reduction procedure presents some challenges due to the identification of heavy ejectiles and the correction of the high-order aberrations through the ray-reconstruction procedure for the much suppressed DCE reaction channel. The ejectiles are identified in atomic number (Z), mass number (A)

IOP Conf. Series: Journal of Physics: Conf. Series **1078** (2018) 012008 doi:10.1088/1742-6596/1078/1/012008

and charge state (q) according to the technique described in refs. [33-34, which provides mass resolution as high as 1/160.

The ray-reconstruction procedure is then applied to the identified set of data, in order to extract the momentum vector at the reaction point of the ejectiles and the absolute cross section. In order to perform an accurate trajectory reconstruction of the measured data, a precise model of the spectrometer response in the specific magnetic setup of the experiment is necessary. The way to test the accuracy of such a model comes from a comparison between the measured phase space parameters at the focal plane and the simulated events for the selected reaction [35].

Once a reliable direct transport map has been obtained, it can be inverted and applied to the measured final coordinates in order to obtain the initial phase space parameters at the target point. These are directly related to the modulus of the ejectile momentum and the scattering angle. Indeed, from the initial vertical $\phi_i$ and horizontal $\theta_i$ angles, the laboratory scattering angle $\theta_{lab}$ is extracted. Then, from the reconstructed momentum, the initial kinetic energy of the ejectiles is deduced. The corresponding $Q$-values, or equivalently the excitation energy $E_x = Q - Q_0$, where $Q_0$ is the ground state to ground state $Q$-value, are finally obtained by a missing mass calculation based on relativistic energy and momentum conservation laws, assuming a binary reaction.

An example of the correlation plot $\theta_{lab}$ versus $E_x$ is shown in Figure 1 for the $^{116}$Cd($^{20}$Ne,$^{20}$O)$^{116}$Sn DCE reaction in the angular range 3° < $\theta_{lab}$ < 14°. The $^{116}$Sn ground state region is visible as vertical and straight locus around $E_x = 0$, even with the low collected yield, as expected since the $E_x$ parameter does not depend on the scattering angle for transitions to the $^{116}$Sn states. The efficiency cut on the bottom of the distribution is due to the presence of the protection screen that limit the FPD acceptance.

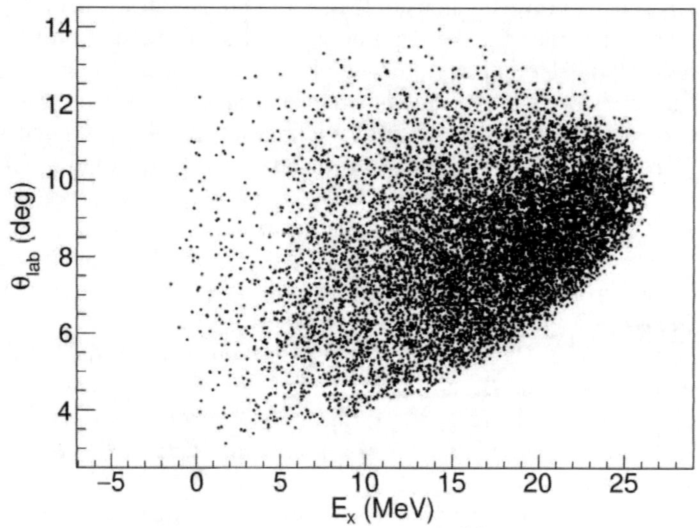


Figure 1 Plot of the reconstructed $\theta_{lab}$ versus the $^{116}$Sn excitation energy ($E_x$) for the $^{116}$Cd($^{20}$Ne,$^{20}$O)$^{116}$Sn reaction at 15 AMeV.

A major achievement of the ray reconstruction technique is the very small systematic error obtained in the horizontal $\theta_i$ (-0.01° ± 0.04°) and vertical $\phi_i$ (-0.05° ± 0.3°) angles, as demonstrated in Ref. [36]. In addition a high resolution is also obtained in $\theta_i$ (0.2°) and $\phi_i$ (0.7°) angles. Regarding the reconstructed momentum modulus, a resolution of 1/1800 with an accuracy better than 1/1600 is obtained for the reaction channels of interest.

When dealing with very rare processes, as the DCE reactions, other important parameters of the experimental measurement are the cross section sensitivity and the rejection factor. In particular, looking at the reconstructed $\theta_{lab}$ versus $E_x$ plot shown in Figure 1, we can see that there are no spurious counts in the region between -7 and -2 MeV. This corresponds to sensitivity better than 1 count within 5 σ confidence level in an energy range of 1 MeV. To estimate the rejection factor in the region of interest for the transition $^{116}$Cd($^{20}$Ne,$^{20}$O)$^{116}$Sn$_{g.s.}$, we estimated the total ejectile flux emerging from the target seen by the solid angle aperture of the spectrometer according to the formula: $N_{reac}^{tot} = N_{targ} N_{beam} \sigma_{reac} \Delta\Omega_{MAGNEX}^{\varepsilon} / 4\pi$. The use of this formula requires the knowledge of: the number of ions/cm$^2$ deduced from the target thickness ($N_{targ}$), the number of incident ions measured by the Faraday Cup ($N_{beam}$), the solid angle seen by MAGNEX taking into account the transport efficiency of the specific setup (in the present case $\Delta\Omega_{MAGNEX}^{\varepsilon}$ ~ 0.041 sr) [37] and the total reaction cross section $\sigma_{reac}$ for the system $^{20}$Ne + $^{116}$Cd ($\sigma_{reac} \cong \pi R^2 \sim 1.8$ b). After the $B\rho$ selection by the dipole magnetic field, the particle identification and the ray-reconstruction technique, we are able to obtain a tiny amount of spurious counts in the DCE region of interest (< 0.25) that corresponds to a rejection factor better than $4 \times 10^{-9}$ in the region of interest.

## 5. Conclusions

The results described in this paper confirm MAGNEX as the ideal instrument for the challenging measurement of heavy-ion induced DCE reactions. Despite the experimental limitations, it allows to measure energy spectra and absolute cross sections for the DCE reaction channel at very forward angles including zero degree.

We measured also other reaction channels (one- and two-proton transfer, one- and two-neutron transfer and SCE), in order to estimate the role of the sequential multi-nucleon transfer routes on the diagonal DCE process. The data reduction and analysis are almost completed and the results will be published soon.

**Acknowledgments**
This project has received funding from the European Research Council (ERC) under the European Union's Horizon 2020 research and innovation programme (grant agreement No 714625).

**References**

[1] Vergados J D, Ejiri H, and Simkovic F 2012 *Rep. Prog. Phys.* **75** 106301
[2] Vogel P 2012 *Jour. of Phys. G: Nucl. and Part. Phys.* **39** 124002
[3] Engel J and Menéndez J (2017) *Rep. Prog. Phys.* **80** 046301
[4] Dell'Oro S, Marcocci S, Viel M and Vissani F 2016 *Adv. High Energy Phys.* **2016** 2162659
[5] Cappuzzello F et al. 2015 *J. Phys.: Conf. Ser.* **630** 12018
[6] Agodi C et al. 2015 *Nucl. Part. Phys. Proc.* **265** 28
[7] Cappuzzello F et al. 2018 *Eur. Phys. J.* A **54** 72
[8] Cavallaro M et al. 2017 *PoS*(BORMIO2017)
[9] Cavallaro M 2017 *PoS*(NEUTEL2017) 031
[10] Cappuzzello F et al. 2016 *Eur. Phys.* J. A **52** 167
[11] Cunsolo A et al. 2007 *Eur. Phys. Jour. ST* **150** 343
[12] Cappuzzello F et al. 2014 *Nucl. Instr. and Meth.* A **763** 314
[13] Cappuzzello F et al. 2015 *Nature Communications* **6** 6743
[14] Cavallaro M et al. 2016 *Phys. Rev.* C **93** 064323
[15] Carbone D et al. 2014 *Phys. Rev.* C **90** 064621
[16] Oliveira J R B 2013 *J. Phys.* G **40** 105101
[17] Carbone D et al. 2011 *J. Phys.: Conf. Ser.* **312** 082016
[18] Laurent H et al. 1993 *Nucl. Instr. and Meth. A* **326** 517
[19] Cavallaro M et al. 2013 *Nucl. Instr. and Meth. A* **700** 65
[20] Cappuzzello F et al. 2015 *Eur. Phys. J.* A **51** 145
[21] Calabretta L et al. 2017 *Modern Physics Letters* A **32** 17
[22] Cortesi M et al. 2017 *Review of Scientific Instruments* **88** 013303
[23] Muoio A et al. 2016 *EPJ Web of Conferences* **117** 10006

[24] Carbone D et al. 2016 *Results in Physics* **6** 863
[25] Carbone D et al. *Nucl. Instr. and Meth.* (in press). DOI:10.1016/j.nima.2017.10.095
[26] De Geronimo G et al. 2013 *IEEE Transactions on Nuclear Science* **60** 2314
[27] Iazzi F et al. 2017 *WIT Transactions on Engineering Sciences* DOI: 10.2495/MC170071
[28] Lazzaro A et al. 2008 *Nucl. Instr. and Methods* A **591** 394
[29] Lazzaro A et al. 2008 *Nucl. Instr. and Methods* A **585** 136
[30] Lazzaro A et al. 2007 *Nucl. Instr. and Methods* A **570** 192
[31] Lazzaro A et al. 2009 *Nucl. Instr. and Methods* A **602** 494
[32] Shima K et al. 1992 *Atom. Data and Nucl. Data Tables* **51** 2
[33] Cappuzzello F et al. 2010 *Nucl. Instr. and Methods* A **621** 419
[34] Calabrese S et al. 2018 *Acta Phys. Pol.* B **49** 275
[35] Carbone D 2015 *Eur. Phys. J Plus* **130** 143
[36] Cappuzzello F et al. 2011 *Nucl. Instr. and Methods* A **638** 74
[37] Cavallaro M et al. 2011 *Nucl. Instr. and Meth.* A **637** 77

IOP Conf. Series: Journal of Physics: Conf. Series **1078** (2018) 012009 doi:10.1088/1742-6596/1078/1/012009

# Characterization of the LEMA isotope separator to measure concentrations of $^{10}$Be from atmospheric filters

**K. De Los Ríos, C. Méndez-García, S. Padilla, C. Solís, E. Chávez, A. Huerta and L. Acosta**

Instituto de Física. Universidad Nacional Autónoma de México. Apartado Postal 20-364 Cd.Mx 01000, Mexico.

E-mail: acosta@fisica.unam.mx

**Abstract.** The understanding and measurement of $^{10}$Be is useful for earth science. Its way of production, deposition and mainly half-life allow phenomenological reconstructions of great interest on this area. In order to characterize the 1 MV isotope separator of the National AMS Laboratory LEMA (Laboratorio de Espectrometría de Masas con Aceleradores) at Mexico City for the measurement of $^{10}$Be concentrations, atmospheric filters with aerosols content were used. The filters were processed with a radio-chemical extraction procedure detailed here. Then, the prepared samples were pressed in cathodes and measured by using the AMS technique, tuning the machine till focus the optics on the beryllium peak. The preliminary statistical analysis of the $^{10}$Be/$^{9}$Be ratios obtained from this characterization is presented in this work. Two statistical methods were used in order to improve the data information. With them, statistical stability behavior is observed on different campaigns of measurements. The results obtained comparing both methods are here discussed.

## 1. Introduction

The Accelerator Mass Spectrometry technique (AMS) was created in the early 80's in order to measure rare and long-lived radioisotopes, which could not be measured using conventional mass spectrometry or decay counting techniques [1]. Among the different radioisotopes that have been measured using the AMS technique, it has been found that $^{10}$Be is of a relevant importance for the development of different sciences [2]. $^{10}$Be is a radioisotope that is produced mainly in the atmosphere, by spallation reactions with $N_2$ and $O_2$. The $^{10}$Be has a half-life of $1.5\times10^6$ years [3], it remains approximately 12 months in the stratosphere, from where it passes to the troposphere staying there for about 2 weeks and is finally deposited on the Earth surface by wet or dry deposition [4]. For this reason, it is considered as a potentially useful tracer to be studied.

In this work, the characterization of the isotope separator of the LEMA IF-UNAM [5] was carried out for the study of $^{10}$Be concentrations in natural samples. The samples chosen were atmospheric filters which were processed making small modifications to the radiochemical procedures established in the literature for the extraction of the radioisotope from this kind of samples [6, 7]. Four different campaigns of measurements were performed at LEMA, calibrating the separator by using reference standards (Labeled as STD**) and blanks (Labeled as BLK**).

IOP Conf. Series: Journal of Physics: Conf. Series **1078** (2018) 012009 doi:10.1088/1742-6596/1078/1/012009

For data analysis the method exploration data analysis (EDA) was used [8]. As a measure of central tendency the median was chosen and as a spread, the median absolute deviation (MAD) [9] was determined. To validate the results, a comparison with the traditional methods was made as well. This kind of analysis allows to determine the $^{10}$Be concentration existing on each processed sample. However in this preliminary results just the $^{10}$Be/$^{9}$Be ratios for the statistical analysis obtained are presented. In order to demonstrate the success of the characterization, a preliminary comparison of some $^{10}$Be concentrations obtained from two different campaigns of measurements, is also presented.

## 2. Methods

### *2.1. Sample extraction and preparation technique.*

The radiochemical extraction was carried out with atmospheric filters collected in 2012 [10]. Mainly the filters of the "Ciudad Universitaria" (Labeled as CU**) and the "Intituto Mexicano del Petroleo" (Labeled as T0**) site were chosen (details regarding samples are well described in [10]). The radiochemical extraction is divided in three main steps: sample preparation, separation and purification of the Be and obtention of the BeO molecule, necessary to produce the negative ions of BeO$^-$ in the ion source to be processed with the AMS technique (see Fig. 1). The radiochemical procedure established here is the result of small modifications (as it is shown in Fig. 1 diagram) of the described procedures found in the literature [6, 7].

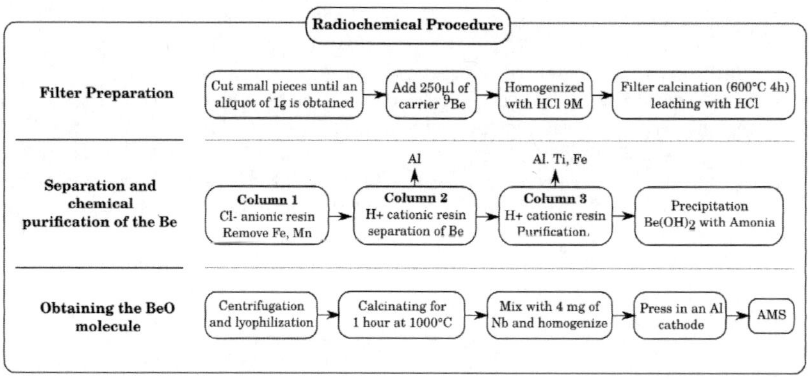


**Figure 1.** Radiochemical procedure adapted for the extraction of $^{10}$Be from atmospheric filters in LEMA.

### *2.2. Measurement in the LEMA isotope separator.*

For the isotope separation it is necessary to take into account the charge, mass and energy properties of the Be isotope, and thus to filter and carry the beam through the beam line, by electric and magnetic fields. The Isotope Separator of LEMA is a Tandetron of 1 MV with a complex optical system coupled. A detailed description of this kind of system can be consulted in [5]. Cathode filled with processed sample are introduced in a carousel of up to 50 positions and then are transported (one by one) to the ion source, where the beam of negative ions is produced. To perform the tuning, the stable isotope beam ($^{9}$Be) is initially focused and the current $I$ is measured in a Faraday cup close to the detection system. Then it is possible to tune the radionuclide of interest ($^{10}$Be) modifying the appropriate parameters in the accelerator. The $^{10}$Be beam is undetectable by the Faraday cups of the accelerator. For this reason it has to be identified and counted event by event, in a gas detector mounted at the end of the line. The

IOP Conf. Series: Journal of Physics: Conf. Series **1078** (2018) 012009 doi:10.1088/1742-6596/1078/1/012009

final filtration of the beam is carried out there. Finally the $\Delta$E-E technique in the gas detector is used in order to observe well discriminated, events of $^{10}$Be. To improve the discrimination between $^{10}$Be and its isobar ($^{10}$B), a 75 nm thick absorber foil is introduced after the electrostatic analyzer. The whole procedure is carried out both for the cathode filled with STD51 (nominal value $^{10}$Be/$^{9}$Be= $2.709 \times 10^{-11}$ [11]) and for the cathode with the sample. It is necessary to introduce a blank cathode (BLK**), with carrier material chemically processed in the same way as the sample, to determine the background level of the technique and to obtain the best possible calibration. The STD ratio in the isotopic separator should be between 8-10% of the nominal STD51 value.

*2.3. Exploratory Data Analysis (EDA).*
In the measurement process a large set of data is provided by the system. The ratios values at the beginning or at the end of the run may change. It is important to establish a statistical method for the data analysis that not be affected by those variations. The method of exploratory data analysis EDA is a favourable way to obtain information from a large data set. The EDA method is robust and resistant for making assumptions regarding the nature of a data set [8]. Within EDA, a reasonably simple complete alternative for the numerical summary is the median, as a measure of central tendency and the median absolute deviation (MAD), as a spread measure. In this case, the MAD is just the transformed value of $y_i(x) = x_i - q_{0.5}$ [9]:

$$MAD = median|y_i(x)| \tag{1}$$

where $x_i$ is referred to each value of the data set and $q_{0.5}$ is the median of the set.

For the statistical analysis of this work, the graphical analysis was made by means of schematic diagrams [8]. This is an extension of the boxes and whiskers diagrams, since this method offers information on the "outliers" behavior, which must be observed for the analysis of this characterization. On this kind of diagram, the box represents the values where the 75% of the data from the statistical distribution is moving, while the whiskers indicate the maximum and minimum value of the data series. Those values that are out of the range established by the whiskers are the so called outlier values. A number of the series is considered an "outlier", when its value is more than 1.5 times the absolute value enclosed by the box, called the inter-quartile range (IQR). Then, the outliers are values so far from the expected behavior of the mean value.

## 3. Results

During the measurement with the AMS technique it is necessary to observe the $\Delta$E-E spectra to be able to determine the validity of the data. This kind of plot shows the separation of the $^{10}$Be and its isobar $^{10}$B, due to the energy loss of the ions in the two anodes of the gas detector. The $\Delta$E-E spectra observed during the tuning process are related to the measurement of: a standard, a blank and a natural sample.

In the first measurement made (April-2017) it was not possible to obtain the desired sequence of spectra since they were far to fulfill the requirements mentioned above. With these results, part of the radio-chemical extraction process pre-established in the literature was changed (as it is described on the Fig. 1). For the second measurements made (September-2017), the spectra were better adjusted to the aforementioned requirements. However, the spectrum of the sample showed a considerable background, evidencing some contaminant in the sample or a calibration problem in the system. For the measurements of November 2017 and February 2018 the radio-chemical technique was established (see Fig. 2) and it was possible to obtain a 10% of the nominal STD51 value, with a beam transmission of 57%. The behavior of the $\Delta$E-E spectra allows to ensure that the radio-chemical extraction was correct and the detected counts effectively were related to $^{10}$Be events and therefore the results of the ratios could be analyzed.

IOP Conf. Series: Journal of Physics: Conf. Series **1078** (2018) 012009 doi:10.1088/1742-6596/1078/1/012009

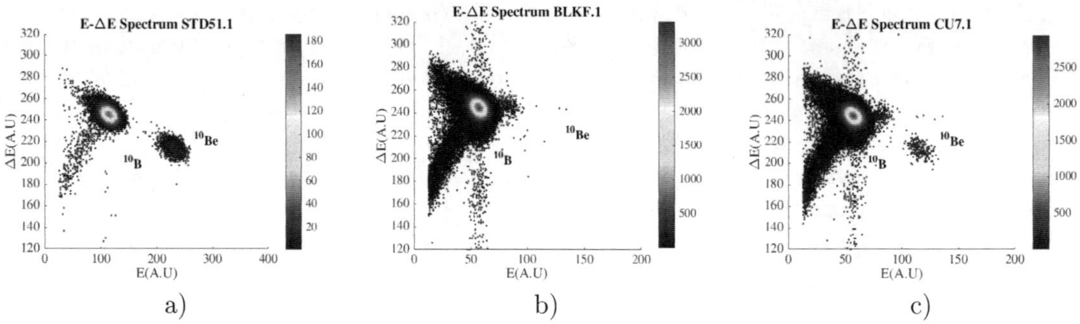


**Figure 2.** E-ΔE spectra observed with the AMS technique for the best campaign of measurements (February 2018) for an acquisition time of 5 minutes: a) STD51 with $^{10}$Be/$^{9}$Be=$1.43\times10^{-12}$ b) Blank processed BLKF with $^{10}$Be/$^{9}$Be=$4.42\times10^{-15}$ and c) Atmospheric sample processed CU7.1 with $^{10}$Be/$^{9}$Be=$2.19\times10^{-13}$.

Since the AMS technique is destructive, the cathodes wear out and for this reason the values at the beginning or at the end of the run may change. To observe this influence on the $^{10}$Be/$^{9}$Be ratios thrown by the system, a dispersion diagram was made showing the outliers using the EDA schematic plot graphic method (see Fig. 3). The mean and the standard error of each cathode involved in the run are also graphed in the Fig. 3. In the data series it is easy to distinguish the standard samples from natural samples and blanks. This behavior is the expected for a good characterization of the isotopic separator. For this reason, the data of Fig. 3 are those considered as the correct and better results, thus completing the characterization of LEMA separator for $^{10}$Be natural samples.

The anomalous samples showed cathodes partially melted during the ionization process, so they do not represent a reliable measurement. In order to make a comparison between the different results, all experimental data coming from anomalous samples were removed from this analysis (purple-shaded in Fig. 3).

## 4. Discussion

### *4.1. Comparison of statistical methods for the different campaigns of measurements.*

Performing the schematic plot for the radio-chemically processed samples (See Fig. 4 a), b) and c)), stability was observed in the dispersion data for the three plots: September-2017, November-2017, and February-2018. A better stability is observed in November-2017 and February-2018 plots, considering that more outliers appear in September-2017. This behavior can be explained, for the case of September-2017, as a problem of contamination in the radiochemical procedure.

For the three campaigns of measurements some of the cathodes were filled with material coming from a same sample. In this context it was possible to develop some systematic comparisons between the data resulting from the different campaigns after the AMS analysis (shaded in pink and green in Fig. 4). These data were identified for each campaign as TO3** and TO4** and then analyzed taking into account both, the traditional statistical method and the EDA method.

Figs. 5 a) and b) show in two different plots, the comparison between results coming from same samples but different campaigns: In Fig. 5 a) the plot shows the results of September-2017 measurements for the natural sample T03, on the right side of the dashed line; In Fig. 5 b) the plot shows the results of November-2017 measurements for the natural sample T04, as well on the right side of the dashed line. In both plots, the February-2018 measurements for some of the same samples (also used on September-2017 (Fig. 5 a)) and November-2017 (Fig. 5 b))

IOP Conf. Series: Journal of Physics: Conf. Series **1078** (2018) 012009 doi:10.1088/1742-6596/1078/1/012009

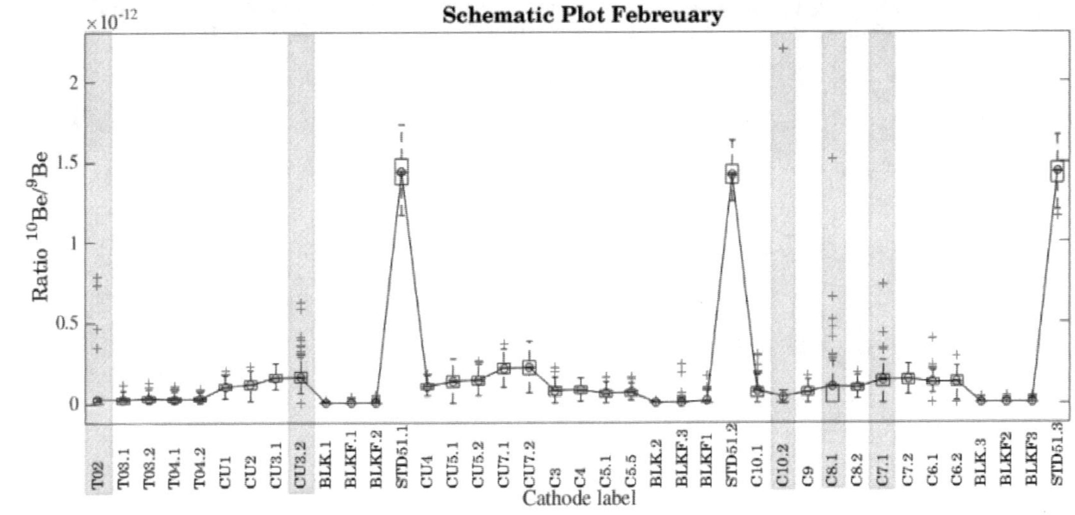


**Figure 3.** Schematic plot for the best campaign of measurements (February-2018), showing $^{10}$Be/$^{9}$Be ratios as a function of the natural samples, standards and blanks. The blue boxes enclose the median (red horizontal line) and the the mean (blue circles) of each statistical distribution. Dashed bars are whiskers representing maximum an minimum values of each series, whereas solid bars are the statistical errors. Red crosses are the outliers values, i.e. values larger than 1.5 times the IQR value (see the text for details). The samples shaded in purple, represent the results with more outliers during the measurement.

campaigns) are shown on the left side of the dashed line.

For the calculation of these results, the normalization was made with respect to the nominal value of STD51. Although the data are from the same sample in both plots of Fig. 5 (T03 and T04), the ratios showed some differences (September, November or February). This difference can be related to the amount of $^{9}$Be carrier added in the radiochemical procedure, then the samples are more or less diluted. Either way, the behavior of the different data series compared in Fig. 5 showed similar conditions observing in all of them a trend comparable taking in to account the uncertainties calculated in every measure. Considering the noise found on the September campaign spectra, such data can be discarded. Then, the best data were those from November and February campaigns, so the parameters related to the sample protocol preparation and the isotope separator tuning were fixed with these last two campaigns. With minor changes in the tuning of the machine from November to February measurements the data were improved, showing a better agreement between both statistical methods used for the analysis (see Fig. 5). This results are saying that February measurements where those when for most of the samples, the spectrometer was more stable on the time of measure.

Anyhow, in order to make a real comparison of the measurements, it is necessary to calculate the final concentrations of $^{10}$Be radioisotope. This has to be made by using the values of standard and blanks (typically around the numbers reported in Fig. 2) and many other parameters related to the chemical preparation for the normalization. On the Table 1, the preliminary concentrations obtained for the sample T04 in November-2017 and February-2018 campaigns are shown. As it can be seen on the Table, this values show a good agreement for both campaigns. Nevertheless, many other tests will be make with the different data campaigns, in order to support the whole technique here presented.

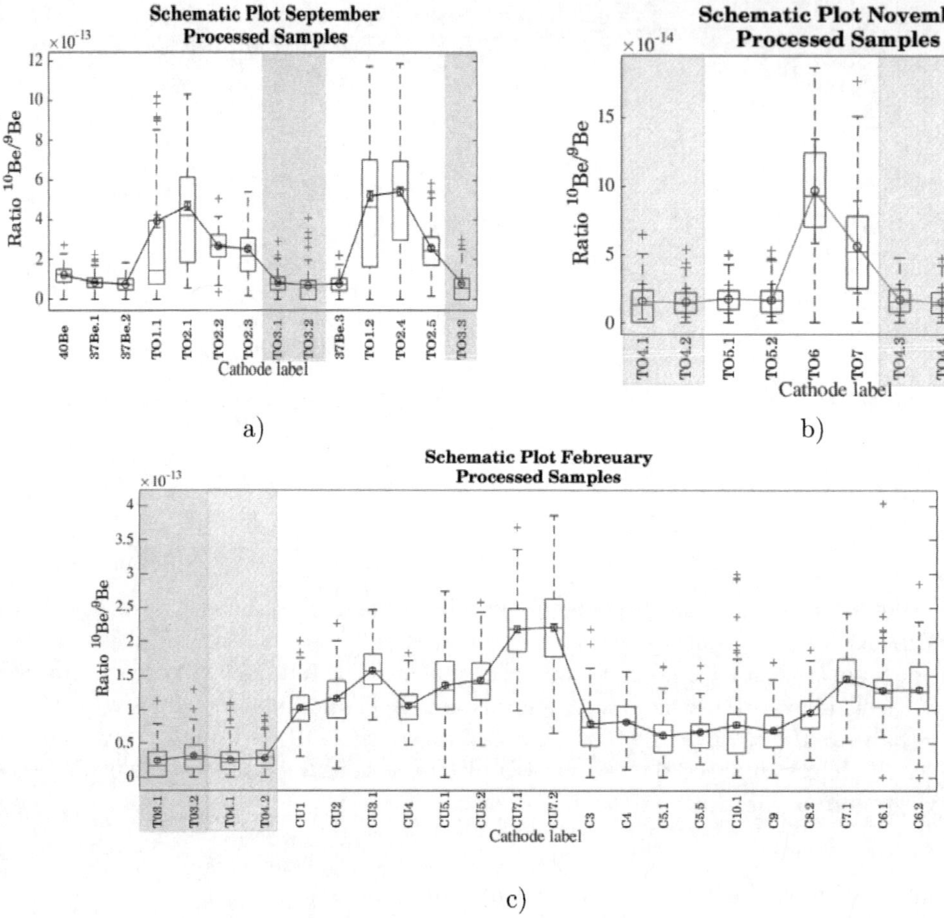


**Figure 4.** The same kind of plot than in Fig. 3, but for different campaigns of measurements: a) September-2017 b) November-2017 and c) February-2018. In this plots, the standards and the blanks ratios measured, as well as those samples with many outliers, were removed. The samples shaded in pink and green on the three plots, denote those samples that can be used to make comparisons: pink, September-2017 with February-2018 and green, November-2017 with February-2018.

| *Sample* | $C_{mean}$ | Error$C_{mean}$ | $C_{median}$ | Error$C_{median}$ |
|---|---|---|---|---|
| T04NO17 | 2.4452e+08 | 2.5749e+07 | 2.6516e+08 | 6.3303e+07 |
| T04FE18 | 2.7126e+08 | 1.1995e+07 | 2.9467e+08 | 1.8501e+07 |

**Table 1.** Preliminary concentrations and their corresponding errors for the sample T04 on November-2017 and February-2018 campaigns. The values obtained with both statistical methods (EDA and conventional) are reported ($C_{median}$ and $C_{mean}$, respectively). The units for the concentrations measured are atoms/g.

## 5. Conclusions

In this work a protocol for sample preparation and its further analysis by using AMS technique were presented. Four measurements were carried out in order to establish the effectiveness of the protocol and the best parameters for the tuning of the LEMA isotope separator.

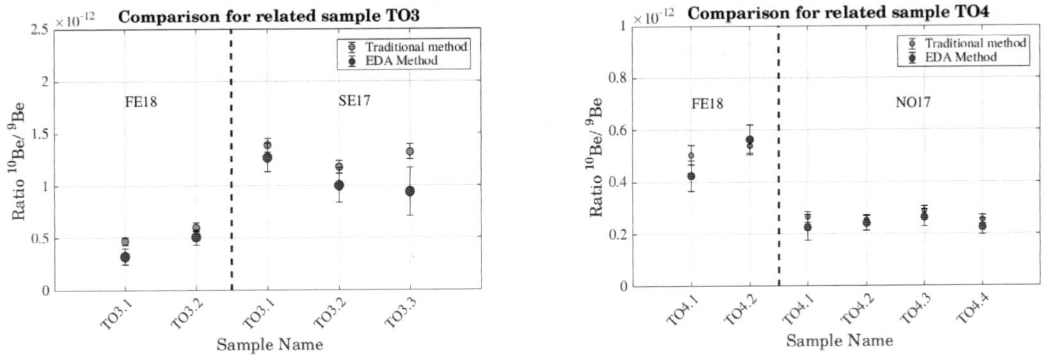


a) b)

**Figure 5.** Comparison of some of the ratios measured in three different campaigns: In the left side of the dashed line in a) and b) are the $^{10}Be/^{9}Be$ ratios for samples from February-2017 campaign. In the right side of the dashed line of both plots are shown some ratios coming from same samples as in February-2017 campaign but for September-2017 campaign in a) and, November-2017 campaign in b). Pink circles are related with a conventional statistical analysis and blue circles with the EDA analysis method. Both data series present their statistical error bars.

In the characterization process carried out, the radiochemical extraction technique is critical to obtain reliable data. The $\Delta$E-E spectra is one of the ways to verify that the extraction procedure is successful. In the measuring process the ratios values at the beginning or at the end of the run may change, considering this, the EDA method allows to determine measures of central tendency and the spread, being unaffected by these variations (if they occur). The method as well is useful to observe the stability of the data as a function of the time in this kind of measurements. Finally, stable and good data from the campaign of February-2018 show the expected behavior, demonstrating that the LEMA isotope separator is ready for the application of AMS technique to the study of $^{10}Be$ concentrations in atmospheric filters with PM10 particles. The agreement found between preliminary concentrations coming from two different campaigns of measurements, demonstrates the good stability of the whole process.

## 6. Acknowledgments.

This work was partially supported by CONACYT 51600, 82692, 123655 Projects and DGAPA-PAPIIT IA101616, IA103218 Projects.

## References

[1] Kutschera W 1983 *Radiocarbon* **25** 677–691
[2] Kutschera W 2013 *International Journal of Mass Spectrometry* **349** 203–218
[3] Chengde S, Beer J, Tungsheng L, Oeschger H, Bonani G, Suter M and Wölfli W 1992 *Earth and Planetary Science Letters* **109** 169–177
[4] Berggren A M 2009 *Influence of solar activity and environment on 10Be in recent natural archives* Ph.D. thesis Acta Universitatis Upsaliensis
[5] Solís C, Chávez-Lomelí E, Ortiz M, Huerta A, Andrade E and Barrios E 2014 *Nuclear Instruments and Methods in Physics Research Section B: Beam Interactions with Materials and Atoms* **331** 233–237
[6] Child D, Elliott G, Mifsud C, Smith A and Fink D 2000 *Nuclear Instruments and Methods in Physics Research Section B: Beam Interactions with Materials and Atoms* **172** 856–860
[7] Domínguez S P 2015 *Medidas de $^{10}$Be y $^{26}$Al en espectrometría de masas con acelerador de baja energía en el Centro Nacional de Aceleradores* Ph.D. thesis Universidad de Sevilla
[8] Tukey J W 1977 *Exploratory data analysis* vol 2 (Reading, Mass.)

[9] Wilks D S 2011 *Statistical methods in the atmospheric sciences* vol 100 (Academic press)
[10] Gómez V, Solís C, Chávez E, Andrade E, Ortiz M, Huerta A, Aragón J, Rodríguez-Ceja M, Martínez M and Ortiz E 2016 *Nuclear Instruments and Methods in Physics Research Section B: Beam Interactions with Materials and Atoms* **371** 365–369
[11] Nishiizumi K, Imamura M, Caffee M W, Southon J R, Finkel R C and McAninch J 2007 *Nuclear Instruments and Methods in Physics Research Section B: Beam Interactions with Materials and Atoms* **258** 403–413

IOP Conf. Series: Journal of Physics: Conf. Series **1078** (2018) 012010 doi:10.1088/1742-6596/1078/1/012010

# First results of the Hoyle-Gamma experiment: study of the excited levels in carbon-12 gamma decay

**F. Favela[1,2], L. Acosta[2,1], L. Auditore[1], A. Camaiani[4], G. Cardella[1], E. De Filippo[1], S. De Luca[5,1], N. Gelli[4], B. Gnoffo[1,8], D.J. Marín-Lámbarri[2], G. Lanzalone[3,6], C. Maiolino[3], N.S. Martorana[3], A. Nannini[4], A. Pagano[1], E.V. Pagano[3], M. Papa[1], S. Pirrone[1], G. Politi[8,1], E. Pollacco[7], L. Quattrocchi[5,1], F. Rizzo[8,3], P. Russotto[3], A. Trifirò[5,1], M. Trimarchi[5,1]**

[1] INFN sezione di Catania, Via Santa Sofia 64 I-95123, Catania, Italy.
[2] Instituto de Física, Universidad Nacional Autónoma de México, 01000, México, D.F., México
[3]INFN LNS, Via Santa Sofía 62 I-95123, Catania, Italy.
[4]INFN sezione di Firenze and Dip. di Fisica Università di Firenze, Italy
[5]Dipartimento di Scienze MIFT, Università di Messina, Messina, Italy.
[6]Facoltà di Ingegneria e Architettura, Università Kore, Enna, Italy.
[7]CEA IRFU Saclay, Gif sur Yvette, France.
[8]Dipartimento di Fisica e Astronomia, Università di Catania, Italy.

E-mail: `francisco.favela@ct.infn.it`

**Abstract.** Preliminary analysis of the Hoyle-Gamma experiment is presented, where simultaneous gamma-particles detection for the $\alpha + {}^{12}\mathrm{C} \rightarrow \alpha + ({}^{12}\mathrm{C}^* + \gamma)$ reaction is used to suppress the background various orders of magnitude. This will allow us to measure very small $\Gamma$ partial widths of ${}^{12}$C. Details of the method, the implementation and CHIMERA software analysis are presented as well as preliminary results.

## 1. Introduction

In previous measurements we have shown the possibility to measure gamma rays with **CHIMERA** detector array looking to the 4.44 decay of ${}^{12}$C [1]. To use such ability and the large solid angle covered by this detector array, we are interested in observing the gamma decay of ${}^{12}$C states above the particle emission threshold, important for astrophysical studies [2], of the Hoyle (at 7.65 MeV) and the 9.6 MeV states.

A multicoincidence technique analysis has been developed in order to obtain a very high rejection fraction of the background, necessary to measure the rare gamma decays of these states. The contemporary detection and identification of particles and gamma rays with the $4\pi$ detector **CHIMERA** [3] guarantee the large efficiency necessary for such a method.

**CHIMERA** has 18 rings in the forward part (covering from $\approx 1°$ to $30°$), and 17 rings in the sphere ($\approx 30°$ to $175°$). Using the azimuthal symmetry, each ring corresponds to a different $\theta$ value and the $\phi$ depends of the specific detector in a particular ring. The target is centered at the center of the sphere. As a convention we put the rings of the forward part in 9 groups with 2 sub-rings each, we differentiate between two sub-rings by tagging them with an "i" or an "e" referring to internal or external sub-ring (for example "8i" refers to ring 8 internal sub-ring). For the sphere we continue the numeration of the rings up to 26, starting from 10. Here we don't need sub-rings tagging and as a convention we add an "S" at the beginning of the group number to obviate that we are referencing a ring on the sphere (for example "S14" refers to ring 14 on the sphere). To refer to any particular telescope in a given ring we put a number next to the ring that contains it, this number goes from zero to the number of detectors of that ring minus one, for example "S11 11" refers to the twelfth telescope on ring "S11". We can also address the telescopes by an absolute number going from 0 ("1i 0") to 1191 ("S26 7"), for example telescope "S11 11" is equivalent to telescope 731. The former way of addressing the telescopes is a ring relative form while the latter is an absolute form.

On this experiment a 64 MeV alpha beam was used to impinge a carbon target (106 $\mu$m/cm$^2$), with the intention of populating various excitation levels of $^{12}$C and measuring the gamma ray angular correlation. On the following sections the suppression method and part of the analysis is shown using the explained conventions.

## 2. Suppression Method

There has been previous work in using **CHIMERA** for gamma rays identification in [1, 4]. Measuring just the gamma rays will not give us any useful information about our system. We need to suppress the background using particle coincidences.

We can start by selecting the rings that should be in kinematic coincidence (using the azimuthal symmetry) and then get to the detector level. Table 1 gives the set of rings that should be in kinematical coincidence for several states in $^{12}$C including the Hoyle state, taking into account that the $^{12}$C was able to punch through the target and the alpha particle was able to reach a CsI detector. The minimum and maximum rings for the residual particle ($^{12}$C) are due to the $\Delta\theta$ on the ejectile ring.

| ground | | | 4.4 | | | 7.65 | | | 9.64 | | |
|---|---|---|---|---|---|---|---|---|---|---|---|
| $\alpha_{ring}$ | $^{12}C_{min}$ | $^{12}C_{max}$ | $\alpha_{ring}$ | $^{12}C_{min}$ | $^{12}C_{max}$ | $\alpha_{ring}$ | $^{12}C_{min}$ | $^{12}C_{max}$ | $\alpha_{ring}$ | $^{12}C_{min}$ | $^{12}C_{max}$ |
| 9i | S15 | S15 | 8e | S14 | S14 | 8i | S14 | S14 | 7e | S14 | S14 |
| 9e | S15 | S15 | 9i | S14 | S14 | 8e | S14 | S14 | 8i | S14 | S14 |
| S10 | S14 | S15 | 9e | S14 | S14 | 9i | S14 | S14 | 8e | S14 | S14 |
| S11 | S13 | S14 | S10 | S13 | S14 | 9e | S14 | S14 | 9i | S14 | S14 |
| S12 | S13 | S13 | S11 | S13 | S13 | S10 | S13 | S14 | 9e | S13 | S14 |
| S13 | S12 | S13 | S12 | S12 | S13 | S11 | S13 | S13 | S10 | S13 | S13 |
| S14 | S11 | S12 | S13 | S12 | S12 | S12 | S12 | S13 | S11 | S12 | S13 |
| S15 | S11 | S11 | S14 | S11 | S12 | S13 | S12 | S12 | S12 | S12 | S12 |
| S16 | S10 | S11 | S15 | S11 | S11 | S14 | S11 | S12 | S13 | S11 | S12 |
| S17 | S10 | S10 | S16 | S10 | S11 | S15 | S11 | S11 | S14 | S11 | S11 |
| S18 | 9e | S10 | S17 | S10 | S10 | S16 | S10 | S11 | S15 | S10 | S11 |
| S19 | 9i | 9e | S18 | 9e | S10 | S17 | S10 | S10 | S16 | S10 | S10 |
| S20 | 8e | 9i | S19 | 9i | 9e | S18 | 9e | S10 | S17 | S10 | S10 |

**Table 1.** Coincidences between rings of **CHIMERA** for the ground state and the first levels of $^{12}$C (considering a 64 MeV alpha beam on $^{12}$C).

As we can see from Table 1, we notice that most of the coincidences are in fact on the sphere, however given the distance from the target to the detector and the size of the detector we end up with a $\Delta\theta = 8°$ (nevertheless, a kinematical coincidence technique can greatly improve the angular resolution [5]), this means that for a specific reaction the alpha energy may vary up to 3 MeV for a given telescope on the sphere.

This energy spread is too large and can be very dangerous for the experiment we performed. We can recover from this and get a good resolution by adding the energies of both particles (alpha and $^{12}C$) and obtaining the Q-value spectrum. This allows to make cuts around specific peaks (corresponding to $^{12}C$ excited levels) and finding the corresponding alphas and carbon particles (see Figure 1, adapted from [6], as an example of a Q-value spectrum that we have to build once the calibration for each of the detectors is ready).

We can further improve the background suppression by kinematically selecting the detectors in coincidence as follows.

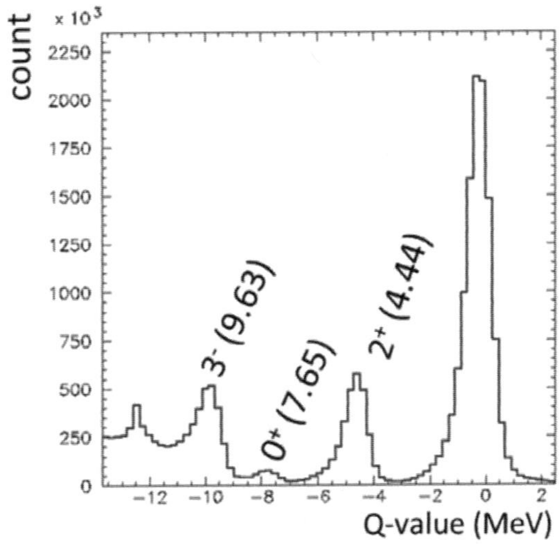


**Figure 1.** Q-value spectrum from a previous experiment [6] with p + $^{12}C$ at 24 MeV.

In accordance to Table 1, in Figure 2 we see that if an alpha particle (ejectile, blue) is detected in ring "8i" (when populating the Hoyle state), then the corresponding $^{12}C$ is detected in ring "S14" at the sphere (red).

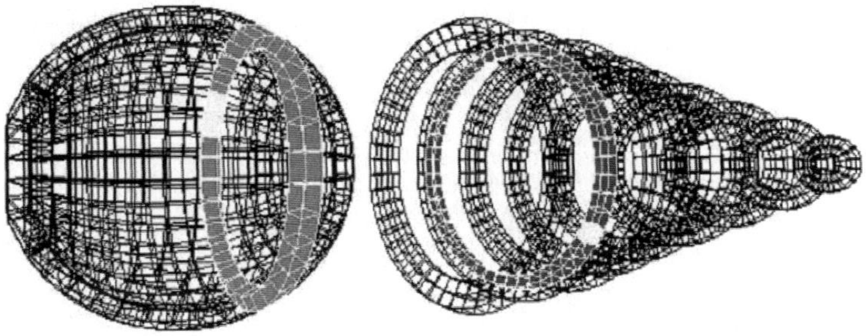

**Figure 2.** Rings in kinematic coincidence. An alpha particle arrives at ring "8i" (blue on the right) and the corresponding $^{12}C$ arrives at "S14" on the sphere (red on the left). Specific detectors in coincidence are shown on yellow.

Figure 2 shows an example of a coincidence from two telescopes, shown in yellow (180° apart

IOP Conf. Series: Journal of Physics: Conf. Series **1078** (2018) 012010 doi:10.1088/1742-6596/1078/1/012010

in $\phi$). Note that for a fixed alpha particle angle, the coincident $^{12}$C recoil could be detected (event by event) in different angularly close telescopes.

In order to properly profit from the $4\pi$ nature of **CHIMERA** we need to take into account every telescope. For each telescope where a $^{12}$C has been detected, inside the specific cut for a specific excitation energy region (on the Q-value spectra) in coincidence with an alpha particle, we have to search for all the telescopes where the corresponding gamma rays on the CsI have been detected.

The better way to extract the angular distribution of emitted gamma rays is to consider as reference the $^{12}$C recoil angle. Therefore we must determine the relative angle between a telescope where the $^{12}$C has been detected and the one where a gamma ray has been detected. The nice symmetry for different rings corresponding to different theta values is broken. Section 2.1 speaks of software tools that have been developed in order to automate this and other tasks.

Figure 3 shows the results of such calculations in a specific case for a detector in ring "8i" (the yellow telescope in the middle where a $^{12}$C is found). In red are marked the detectors that have a relative angle (with respect to the yellow one) close to 0°, in green the ones close to 90° and in blue the ones close to 180°. The zero degree line is defined from the target to the center of the telescope where the $^{12}$C has been detected.

We can take into account for the scarce statistics (specially expected for the 9.6 MeV state) by adding all the gamma histograms that come from the coincidences with a corresponding $^{12}$C. There should be 2 gamma rays for each $^{12}$C (for the aforementioned levels above the 4.4 MeV), making in total a 4 fold particle coincidence increasing even more the background suppression.

A fraction of the gamma rays from the 9.6 MeV level could go directly to the ground state, making some of the coincidences 3 fold.

### *2.1. Software Development*

The analysis is greatly simplified and sped up by the developed software named **alchemist** (previously named binCor). It uses isonav [7] as a kinematic engine. You can find alchemist under the isonavScripts folder provided at the link on the reference.

It receives as input a file with the reaction details, beam energy projectile isotope, residual isotope, excitation energy of the residual particle and target thickness and depending on the selected option it will output a table with the corresponding rings in coincidence, the energies of the particles before and, if chosen, the energy loss after the silicon detector. It also includes time of flight and specific detector operations. Table 1 as well Figures 2 and 3 where made with the help of this software. It also provides general information of **CHIMERA** and example generation.

## 3. Experimental procedure

The experiment was performed at the "**L**aboratori **N**azionali del **S**ud" (**INFN-LNS** Catania, Italy). We used a 64 MeV alpha beam impinging on a natural carbon target ($\approx 98.9\%$ $^{12}$C and $\approx 1.1\%$ $^{13}$C for our intended purposes).

The used natural carbon targets had densities of 50, 69 and 106 $\mu$m/cm$^2$. Having an approximate abundance in the order of one percent on the natural carbon, $^{13}$C may play an important source of spurious reactions. We take this into account by using a pure $^{13}$C target as a control with 50 $\mu$m/cm$^2$ as density.

A hybrid acquisition system has been implemented (VME and GET [8]) as part of a continuous upgrade in order to satisfy our experimental needs.

The system was set to trigger on multiplicity $\geq 2$ on the silicons; of course the gamma rays arrive first but the rise-time of the CsI is so large ($\approx 1\mu$ sec) that the charged particles signals catch up and get in time for the triggering.

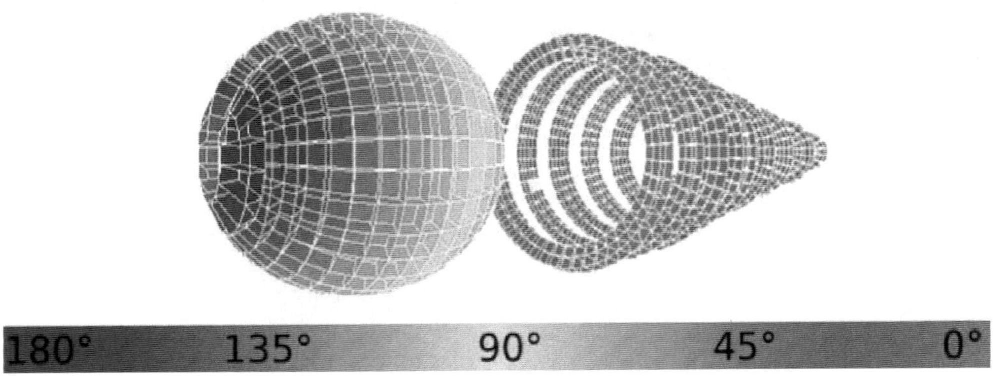


**Figure 3.** Relative angle of all the telescopes in **CHIMERA** with respect to the yellow one (in the middle).

## 4. Preliminary results

Figure 4 shows a Si vs fast (CsI) spectra on the 731 telescope ("S11 11"). We can clearly see the regions of p, d, t, the $^3$He and the alpha. We are also able to fairly distinguish the different alphas corresponding to the different excitation levels, given the angular coverage of the detector, the energies are not well resolved as in [9] but, as it has been previously mentioned, the Q-value spectra will help us overcome any overlaps between levels.

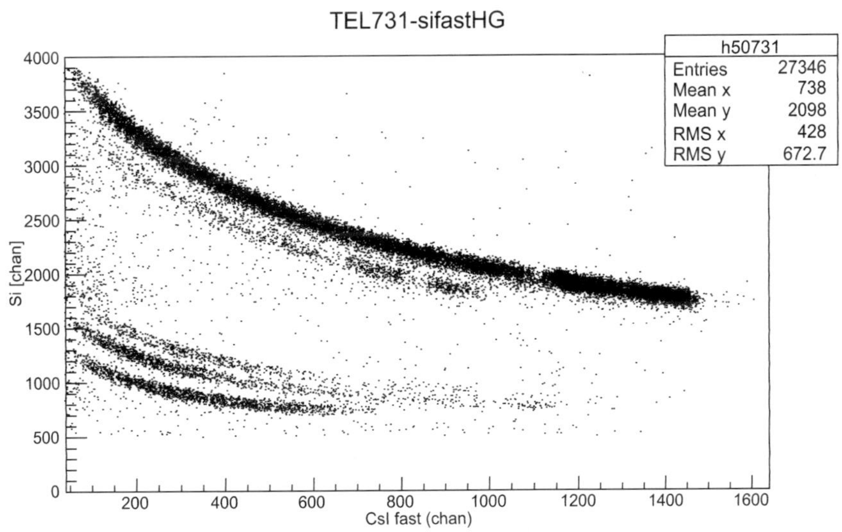


**Figure 4.** Si-fast spectra of the 731 telescope ("S11 11"). Alpha and $^3$He particles are seen at the top while the 3 regions at the bottom correspond to p, d, t (protons, deuterons and tritons).

If we choose a specific energy to study (for example the Hoyle state) and select the alpha particles on a particular telescope (in this case 731 or "S11 11") we could also locate the corresponding detector where the residual $^{12}$C particle should be found. We know from Table 1 that for this reaction "S11" (ejectile) ring is in kinematic coincidence with ring "S13" (residual).

IOP Conf. Series: Journal of Physics: Conf. Series **1078** (2018) 012010 doi:10.1088/1742-6596/1078/1/012010

More specifically "S11 11" (731) and "S13 27" (811) should be in kinematic coincidence for the selected reaction (in a similar manner as in Figure 2).

Figure 5 shows a comparison of two dE vs tof plots of telescopes that are close by in the same ring, 811 ("S13 27") and 808 ("S13 24"). As aforementioned, given the telescope 731 the kinematic coincidence points to telescope 811 (Figure 5.a shows the carbons), that means that there cannot be carbons in other telescopes even if they are close by, this is demonstrated in Figure 5.b.

However alpha particles were seen, as expected from the fragmentation of the $^{12}C^{*}$, in all detectors. Figures 5.a and 5.b show these.

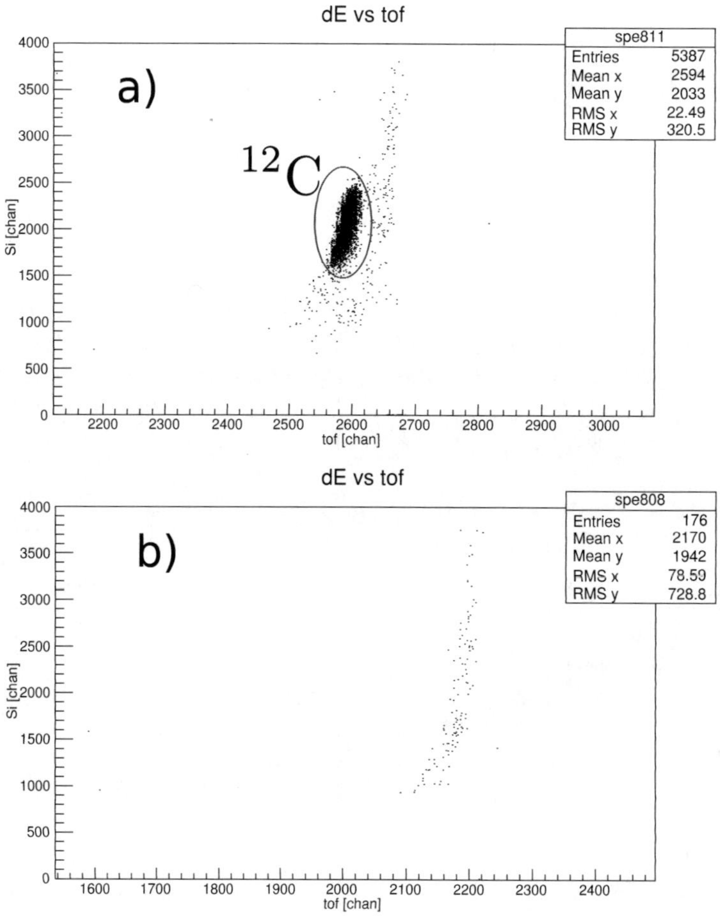


**Figure 5.** dE vs tof for detectors 811 (in kinematic coincidence) and 808 (not in coincidence). Alpha particles from $^{12}C^{*}$ can be seen in both plots.

In accordance to Table 1 if the ejectile reached a detector in ring S13 then, for both the ground and the 4.44 levels, the carbon in kinematic coincidence should be in ring S12. This can readely be shown in Figure 6 where the carbon energy on the front silicon detector is shown for telescope 774 (S12 22) versus the energy (in channels) deposited by the ejectile (alpha) particle from telescope 790 (S13 6). The region enclosed by the polygon is the ground state whereas the region to the left corresponds to the 4.44 level. The energy spread is due to the angular span of the telescopes.

It is important to mention that given the high number of detectors involved in **CHIMERA**, not all of the detectors were expected to work properly during all the runs. Therefore, in order

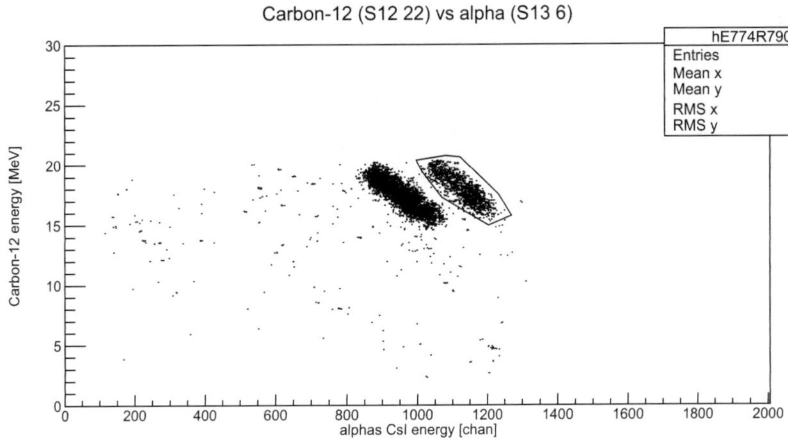


**Figure 6.** Carbon energy [MeV] in Si (telescope 790) vs alpha energy [channels] in CsI. (telescope 774)

to protect the trustworthiness of the data, we simply avoid using these detectors at all (at least in this initial analysis) and we will take this into account in the efficiency calculation.

## 5. Conclusions

We showed how the granularity challenge of the high angular coverage for each of the telescopes in the sphere can be overcomed by the Q-value plot (still pending for this experiment). And how subsequent background suppressions can be performed by means of kinematic coincidences, while at the same time take into account, telescope by telescope, the angular distribution of the gamma rays.

Up to a 4 fold coincidence can be obtained by simultaneous measurements of charged particles with gamma rays, making the technique a good background suppression tool.

We were able to acquire the data with a hybrid system successfully and performed basic analysis allowing us to confirm the reliability of the data in order to make the final analysis.

The analysis software seems to work properly and will help to speed up the process. It is under continuous improvement, future upgrades will provide more options and a better graphic display. It allows to immediately start handling similar types of experiments, where a binary reaction is studied with **CHIMERA**.

## References

[1] Cardella G *et al.* 2015 *Nuclear Instruments and Methods in Physics Research Section A: Accelerators, Spectrometers, Detectors and Associated Equipment* **799** 64 – 69 URL `http://www.sciencedirect.com/science/article/pii/S0168900215008839`

[2] Herwig F, Austin S M and Lattanzio J C 2006 *Phys. Rev. C* **73**(2) 025802 URL `https://link.aps.org/doi/10.1103/PhysRevC.73.025802`

[3] Pagano A *et al.* 2004 *Nuclear Physics A* **734** 504 – 511 URL `http://www.sciencedirect.com/science/article/pii/S0375947404001137`

[4] Martorana N 2018 *submitted to Phys. Lett. B*

[5] Acosta L *et al.* 2013 *Nuclear Instruments and Methods in Physics Research Section A: Accelerators, Spectrometers, Detectors and Associated Equipment* **715** 56 – 61 URL `http://www.sciencedirect.com/science/article/pii/S0168900213003161`

[6] Cardella G *et al.* 2017 *EPJ Web Conf.* **165** 01009 URL `https://doi.org/10.1051/epjconf/201716501009`

[7] Favela F 2018 isonav
https://github.com/ffavela/isonav
[8] Pollacco E *et al.* 2018 *Nuclear Instruments and Methods in Physics Research Section A: Accelerators, Spectrometers, Detectors and Associated Equipment* **887** 81 – 93 URL
http://www.sciencedirect.com/science/article/pii/S0168900218300342
[9] Favela F *et al.* 2015 *Phys. Rev. ST Accel. Beams* **18**(12) 123502 URL
https://link.aps.org/doi/10.1103/PhysRevSTAB.18.123502

IOP Conf. Series: Journal of Physics: Conf. Series **1078** (2018) 012011 doi:10.1088/1742-6596/1078/1/012011

# Nuclear Astrophysics With Gamma-Beams

**Moshe Gai**

LNS at Avery Point, University of Connecticut, Groton, CT 06340, USA

E-mail: moshe.gai@uconn.edu http://astro.uconn.edu

**Abstract.** Gamma-Beams now available at the HI$\gamma$S facility of the TUNL at Duke University and soon to be available at the Extreme Light Infrastructure - Nuclear Physics (ELI-NP) in Magurele, near Bucharest, Romania, present scientific opportunities for progress in Nuclear Physics and Nuclear Astrophysics. In particular the use of Time Projection Chamber (TPC) detectors in these facilities will address the large confusion of conflicting data on the $^{12}$C$(\alpha,\gamma)$ reaction and promise to yield progress on this four decades old problem.

## 1. Introduction: The $^{12}$C$(\alpha,\gamma)$ Reaction

Over the last four decades conflicting data plagued our attempts to deduce the cross section of the $^{12}$C$(\alpha,\gamma)$ reaction at low energies and did not allow an accurate extrapolation of the astrophysical s-factor to stellar energies. In particular conflicting data did not allow us to chose between the high value ($\sim$80 keVb) and the low value ($\sim$10 keVb) solutions of the E1 s-factor at stellar energies. The so called "cascade" s-factors were deduced with large uncertainty, as large as a factor of 25. And the E1-E2 mixing phase angle ($\phi_{12}$) was shown to conflict with unitarity [1]. Recent modern measurement of $S_{E1}$ and $S_{E2}$ at Stuttgart, were demonstrated [1] to have error bars which are considerably larger than quoted by the authors [2, 3]. In spite of the little progress in measurements of the cross section of the $^{12}$C$(\alpha,\gamma)$ reaction, several recent R-Matrix global analyses claim to achieve accuracies of the total s-factor (E1 + E2 + cascade) between 4.5% and 12%. In this paper we first examine one such recently published claim [4] and we point out to the use of gamma-beams to allow for future progress in the field.

The data used in a recent R-Matrix global analysis of the $^{12}$C$(\alpha,\gamma)$ reaction [4] are shown on the left of Fig. 1. We note that in spite of the poor quality of all eight angular distributions measured below $E_\alpha$ = 2.0 MeV, the authors still claim to be within reach of extracting the total s-factor (E1 + E2 + cascade) with accuracy nearing 10%. Furthermore, in the same figure we show the published analysis of the very same angular distribution measured at $E_L$ = 1.79 MeV. Other angular distributions used in [4] were also analyzed in [1] and the extracted error bars are shown to be very large, similar to the error demonstrated at $E_L$ = 1.79 MeV. These analyses (e.g. of the angular distribution measured at $E_L$ = 1.79 MeV) [1] demonstrate that the E2/E1 cross sections cannot be determined with sufficient accuracy (e.g. better than a factor of 6 at $E_L$ = 1.79 MeV!), casting strong doubt on the authors [4] claim of reaching a 10% accuracy in extracting the astrophysical cross section factor of the $^{12}$C$(\alpha,\gamma)$ reaction.

## 2. Gamma-Ray Fcailities: The HI$\gamma$S and ELI-NP

A HI$\gamma$S gamma-ray facility has been operating by TUNL at Duke University [5] over the last two decades and a new gamma-ray facility is constructed by the EU at Magurele near Bucharest

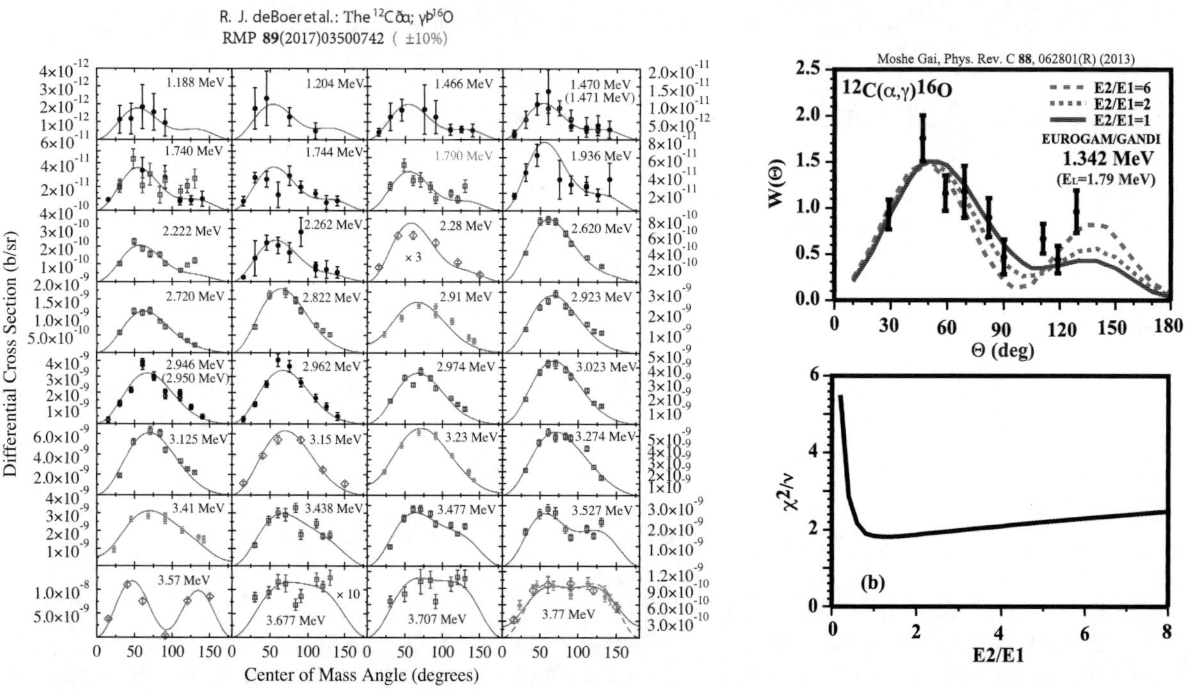


**Figure 1.** The measured angular distributions of the $^{12}C(\alpha,\gamma)$ reaction used in Ref. [4], and analyzed in [1]; e.g. at $E_L$ = 1.79 MeV, as hown on the right.

.

Romania [6]. These gamma-ray facilities promise to be essential for progress of measurements in nuclear astrophysics and in particular to resolve the long standing confusion of measurement of the cross section of the $^{12}C(\alpha,\gamma)$ reaction.

Active target TPC detectors are a perfect match to gamma-ray ray beams and specifically they allow measurements of complete and detailed angular distributions with unprecedented details since the event can be binned in small size bins (e.g. of 4°) as allowed by the statistics. Such detailed angular distributions measured at the large backward angle range are essential for accurate measurements of the E1 and E2 contributions as shown in Fig. 1.

A schematic diagram of the HI$\gamma$S O-TPC [7] is shown in Fig. 2 and a rendering of the ELITPC [8] is shown in Fig. 3. While the O-TPC used the opto-electronic chain shown in Fig. 2 for readout of events, the ELITPC will use the electronic u-v-w readout placed on a single multi-layered board as shown in Fig. 3. These detectors are intended among other projects for measuring the $^{12}C(\alpha,\gamma)$ reaction with the TPCs operating with 100 torr $CO_2$ gas [9].

In Fig. 4 we show the anticipated E2 and E1 cross section factors that will be measured at the HI$\gamma$S (down to 1.7 MeV) and at the ELI-NP (down to 1.1 MeV).

## 3. Conclusions

R-Matrix global analyses cannot resolve the conflicting measurements on the $^{12}C(\alpha,\gamma)$ reaction and a new and bold approach is needed to measure the cross sections at very low energies. This goal quite possibly may be achieved with gamma-ray beams in the newly constructed ELI-NP

IOP Conf. Series: Journal of Physics: Conf. Series **1078** (2018) 012011 doi:10.1088/1742-6596/1078/1/012011

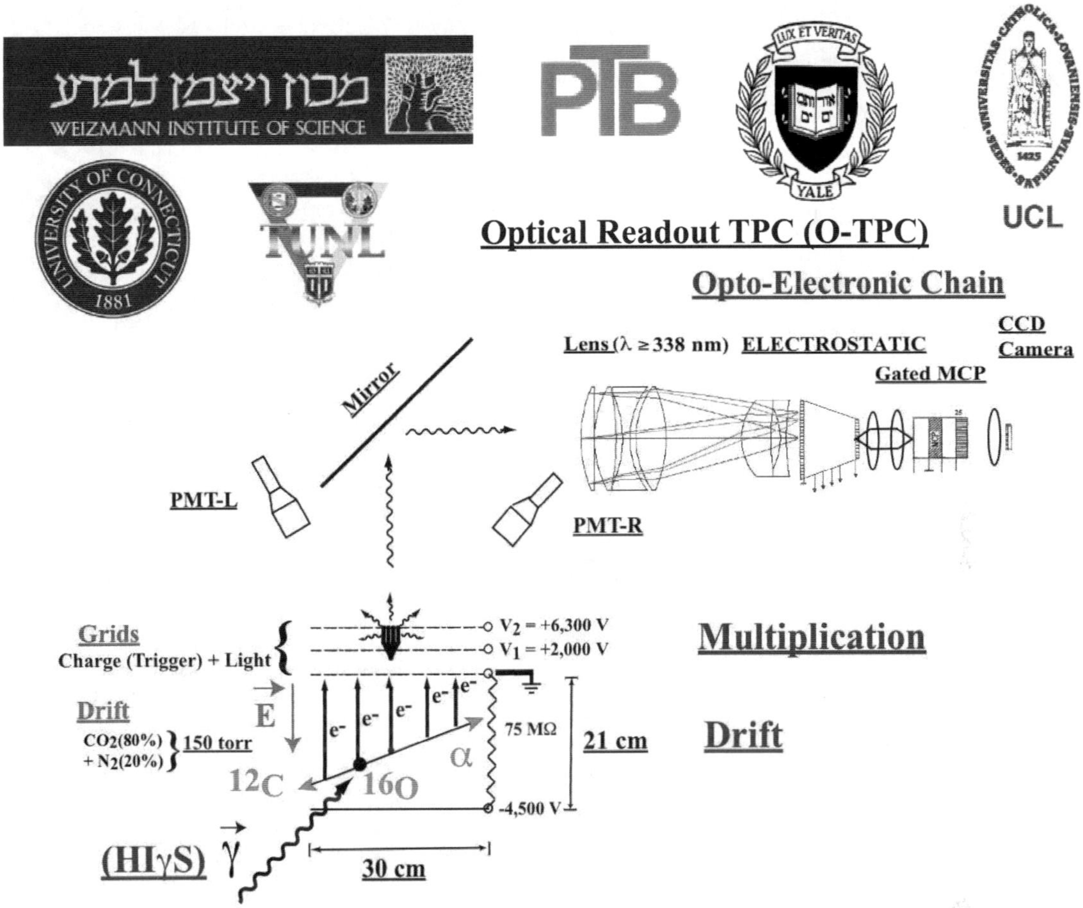


**Figure 2.** A schematic diagram of the HI$\gamma$S O-TPC [7].

and the currently operating HI$\gamma$S facility.

## 4. Acknowledgements

The material presented in this paper is based upon work supported by the U.S. Department of Energy, Office of Science, Office of Nuclear Physics, under Award Number DE-FG02-94ER40870.

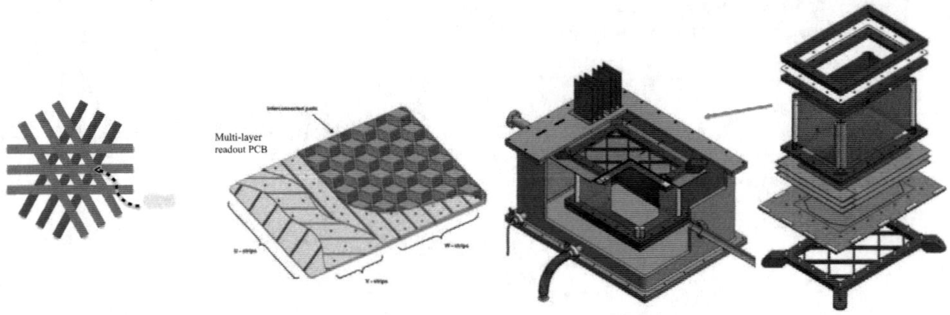

Figure 3. A rendering of the ELITPC with the u-v-w multilayer board readout [8].

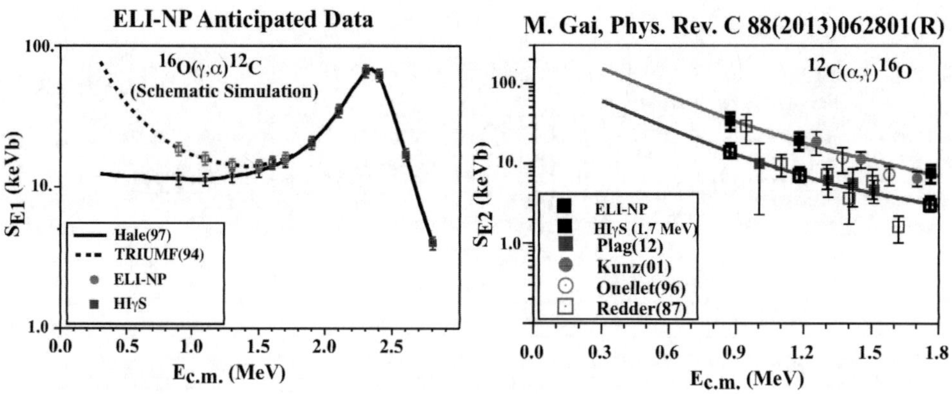


**Figure 4.** Anticipated results from measurements with the HI$\gamma$S O-TPC [7] and all ELI-NP ELITPC [8] detectors.

## References

[1] Moshe Gai, Phys. Rev. C **88**, 062801(R) (2013).
[2] J.W. Hammer *et al.*, Nucl. Phys. **A752**, 514c (2005).
[3] M. Assuncao *et al.*, Phys. Rev. C **73**, 055801 (2006).
[4] R.J. deBoer *et al.*, Rev. Mod. Phys. **89**, 035007 (2017)
[5] H.R. Weller, M.W. Ahmed, H.Gao, W. Tornow, Y. Wu, M. Gai, R. Miskimen, Prog. Part. Nucl. Phys. **62**, 257 (2009).
[6] D. Filipescu, A. Anzalone, D.L. Balabanski, S. S. Belyshev, F. Camera, M. La Cognata, P. Constantin, L. Csige, P. V. Cuong, M. Cwiok, V. Derya, W. Dominik, M. Gai, S. Gales, I. Gheorghe, B.S. Ishkhanov,

A. Krasznahorkay, A. A. Kuznetsov, C. Mazzocchi, V.N. Orlin, N. Pietralla, M. Sin, K.A. Stopani, O. Tesileanu, C.A. Ur, I. Ursu, H. Utsunomiya, V.V. Varlamov, H.R. Weller, N.V. Zamfir, and A. Zilges, Eur. Phys. J. **A51**, 185 (2015).

[7] M. Gai, M.W. Ahmed, S.C. Stave, W.R. Zimmerman, A. Breskin, B. Bromberger, R. Chechik, V. Dangendorf, Th. Delbar, R.H. France III, S.S. Henshaw, T.J. Kading, P.P. Martel, J.E.R. McDonald, P.-N. Seo, K. Tittelmeier, H.R. Weller, and A.H. Young., Jour. Instr. **5**, 12004 (2010).

[8] M. Cwiok, M. Bieda, J.S. Biha?owicz, W. Dominik, Z. Janas, L. Janiak, J. Manczak, T. Matulewicz, C. Mazzocchi, M. Pfuetzner, P. Podlaski, S. Sharma, M. Zaremba, D. Balabanski, A. Bey, D.G. Ghita, O. Tesileanu, M. Gai, Act. Phys. Pol. **49**, 1001 (2018).

[9] O. Tesileanu, M. Gai, A. Anzalone, C. Balan, J.S. Bihalowicz, M. Cwiok, W. Dominik, S. Gales, D.G. Ghita, Z. Janas, D.P. Kendellen, M. La Cognata, C. Matei, K. Mikszuta, C. Petcu, M. Pfuetzner, T. Matulewicz, C. Mazzocchi, C. Spitaleri, Rom. Rep. Phys. **68**, S699 (2016).

IOP Conf. Series: Journal of Physics: Conf. Series **1078** (2018) 012012 doi:10.1088/1742-6596/1078/1/012012

# Weak processes in astrophysical nucleosynthesis

**B. Knight, L. Graham, and O. L. Caballero**
Department of Physics, University of Guelph, Guelph, ON, Canada
E-mail: ocaballe@uoguelph.ca

**Abstract.** Neutron capture and neutrino absorption are of fundamental importance in the synthesis of elements. Both processes can be affected by the extreme conditions of the matter undergoing nucleosynthesis. In this work, we present results of neutron capture rates on $^{46}$Mg and several Sn isopotes. We find that the high degeneracy of neutron matter, a result of the large pressures found in a neutron star, changes the captures rates by orders of magnitude compared to those of non-degenerate neutron matter. We also study the neutrino opacity above the symmetry axis of a black hole accretion disk. We find that there is a region, where matter circulates around the black hole in stable orbits, with potentially low neutrino opacity. This motivates further studies of neutrino emission in the vicinity of a black hole.

## 1. Introduction

Up to this point, the origins of elements throughout the nuclear chart are not well understood. Theoretically speaking, both uncertainties on the astrophysical site and on nuclear reactions leave many open questions. The recent observation of the merger of two neutron stars [1] and the correlated kilonova observation [2] have confirmed this environment as a site for the synthesis of neutron-rich elements. Similarly, X-ray observations (for a compilation see e.g. [3]) have pointed at accretion disks around neutron stars as a possible site for the synthesis of proton-rich nuclei. As more observational data become available, new insights will light up our path to the understanding of the nuclesoynthesis puzzle.

The synthesis of elements depends on diverse nuclear reactions at extreme thermodynamic conditions, with occurrences across hundreds of unstable nuclei in different mass regions of the nuclear chart. Experimental facilities around the world pursue exciting efforts to extend their limits and reach conditions and/or nuclides similar to those found in stellar environments. Among other reactions, neutrino captures on protons and neutrons, and neutron captures on nuclei, play a key role. Neutrino absorption is important in the setting of the electron fraction, determining how neutron-rich the material undergoing nucleosynthesis is. On the other hand, neutron captures naturally occur in neutron-rich environments like Supernovae and mergers. However, neutron captures could also be important in accreting neutron stars, where the highly dense neutron-rich crust material can react with the companion star's matter or the ashes of nucleosynthetic products formed in outter layers [4, 5].

In this contribution, we discus the role of neutron degeneracy on neutron captures, and some concepts on the of neutrino scattering in accretion disks around black holes.

IOP Conf. Series: Journal of Physics: Conf. Series **1078** (2018) 012012 doi:10.1088/1742-6596/1078/1/012012

## 2. Neutron degeneracy in accreting neutron stars

When a neutron star is in the presence of a companion star, stellar atoms such as hydrogen and helium are pulled into the neutron star's envelope. This hot matter is captured, heated, and predicted to undergo thermonuclear burning, producing X-ray bursts and Superbursts [6]. Due to a neutron star's immense gravitational strength, incoming matter reaches velocities fast enough to overcome radioactive decay. Proton captures on nuclei then occur faster than beta decay. This is known as a rapid-proton capture process (rp-process). The composition of the rp-process ashes varies according to the accretion conditions [5, 7]. These ashes can be buried further in the crust where neutrons are abundant and undergo neutron captures. As densities inside the neutron star can reach massive values, comparable to that of the nuclear density [8], neutrons are subject to massive pressures that are responsible for very strong quantum effects governed by the Pauli exclusion principle, i.e they are degenerate. This degeneracy will affect the rates at which neutrons are captured. Thus, a logical improvement for nucleosynthesis studies in this environment is to evaluate neutron capture rates with neutrons obeying a Fermi-Dirac distribution instead of a Maxwell-Boltzmann distribution as is customary in stellar nucleosynthesis studies (see e.g. [9, 10]).

The thermodynamical conditions of a neutron star are not reproducible on earth, and so capture rates must be evaluated theoretically. Neutron degeneracy further complicates this task. Specifically, cross sections would be needed for a wide range of mass numbers, the vast majority of which have not been experimentally measured. Thus, one needs to rely on theoretical cross sections. Additionally, at the temperatures and chemical potentials relevant to a neutron star's crust, the behaviour of the Fermi-Dirac distribution can present abrupt changes.

Shternin et al [11] suggested the use of a power-law behaviour for the relevant cross sections and use properties of the Fermi-Dirac integrals to perform an analytic evaluation of neutron captures on some nuclei. These authors compared their analytical results to numerical integration and found that the results agreed well. However, the results can depend on the energetics of the reaction (and so on the isotopes). In this work, we numerically evaluate the Fermi-Dirac integrals and use theoretical cross sections obtained in the framework of the statistical Hauser-Feshbach model from the specialized nuclear reaction code TALYS [12]. In this context, the total cross section of the reaction $X + n \to Y + \gamma$, in local thermal equilibrium is $\sigma*$, where the asterisk, $*$, highlights the inclusion of thermally excited nuclear levels.

The physical reaction rates are realized by the average $\langle \sigma^* v \rangle$ of the total cross section [11]. This statistical average is calculated by weighting the cross section, $\sigma^*$, over the velocity distribution functions of the target and incident particles (see e.g. [13]). By use of a unitary transformation, to the center of mass and relative coordinates, and integrating out the center of mass term we get

$$\langle \sigma^* v \rangle = \sqrt{\frac{2}{\mu}} \frac{1}{N} \int_0^\infty E\sigma^*(E) f(E) dE,$$

where $N$ is the normalization factor given by

$$N = \int_0^\infty \sqrt{E} f(E) dE,$$

and the collision energy is $E = \mu v^2/2$, where $\mu$ is the reduced mass. Assuming a non-relativistic collision then the reduced mass, the average velocity, $v$, and the distribution function, $f(E)$ correspond to those of neutrons. The distribution function, $f(E)$, will be either the Maxwell-Boltzmann distribution, $f(E)_{MB} = e^{-E/T}$, where $T$ is the temperature in MeV, or the Fermi-Dirac distribution, $f(E)_{FD} = [1 + e^{(E-\mu_n)/T}]^{-1}$, where $\mu_n$ is the neutron chemical potential. To measure the deviation from the Maxwell-Boltzmann averaged rates caused by the degeneracy,

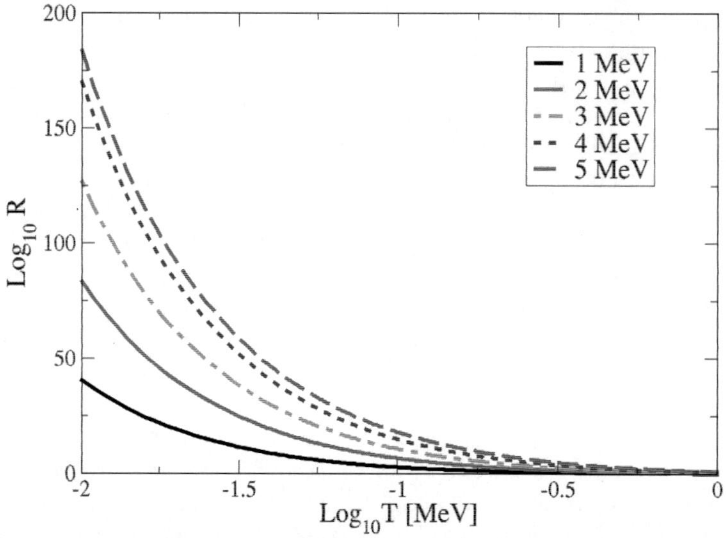


**Figure 1.** The R ratio for the $^{46}$Mg$(n,\gamma)^{47}$Mg reaction at neutron chemical potential $\mu_n$=1, 2, 3, 4 and 5 MeV

we follow the methods in [11] and define a ratio

$$R = \frac{\langle \sigma^* v \rangle_{FD}}{\langle \sigma^* v \rangle_{MB}}.$$

The ratio, $R$, is the cornerstone of this investigation and in principle should be calculated over a range of temperature and chemical potential for different nuclei predicted to be in a neutron star. Large values of $R$ indicate a significant deviation in capture rate, and will be further investigated. Plasma effects, as reported in [11], can also contribute to $R$, but for now we will focus on the effects of degeneracy.

### *2.1. Preliminary results*

Figure 1 shows our $R$ numerical results the reaction $^{46}$Mg$(n,\gamma)^{47}$Mg. These results reproduce those of Shternin et al [11] who used a power law cross section for the same reaction, which agreed with experimental results. The dependence on the chemical potential is apparent. Larger chemical potentials result in a difference of many orders of magnitude, which become even more prominent at lower temperatures.

Figure 2 shows the ratio $R$ for different Sn isotopes ($^{126,132,138,140,165}$Sn) with a chemical potential $\mu_n$= 5MeV. The trend in this case is to have larger changes of neutron captures as the isotope's neutron number, $N$, increases. However, there is a drop in $R$ when $N$ is quite large, as can be seen when comparing $R$ for $^{132}$Sn with $^{165}$Sn. Further studies are needed to understand this behaviour. For now, it is worth noting that in the two cases shown here, neutron degeneracy can change neutron capture rates by orders of magnitude. We aim to investigate many of the exotic nuclei found inside a neutron star's crust.

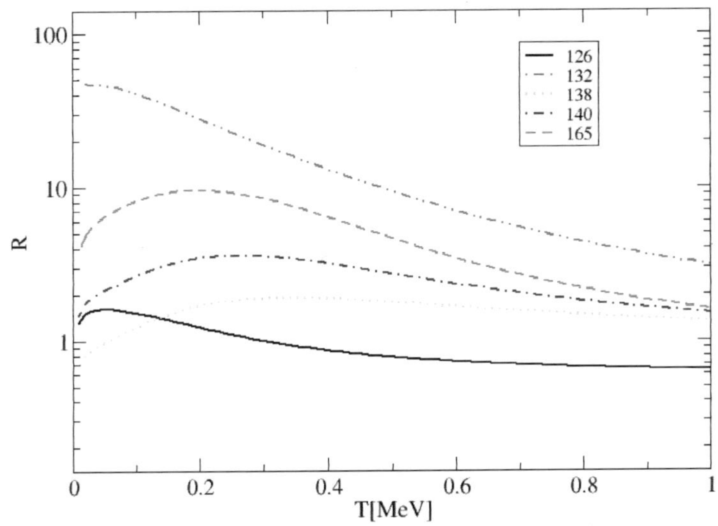


**Figure 2.** The R ratio for neutron captures on Sn isotopes at a neutron chemical potential $\mu_n$=5 MeV. The lines indicate the isotope's mass number, $A$.

## 3. Neutrino opacities and surfaces in black hole-neutron star mergers

There are only a few possible places where the synthesis of heavy neutron-rich elements can occur. This is because their formation is reliant on the rapid neutron capture process (r-process) which requires a high energy environment. There are few sites known that yield such high energies, one such being the hot outflows ejected from accretion disks. Such disks can be formed during a black-hole neutron star (BH-NS) merger and emit a large amount of neutrinos. The matter of these disks is highly dense, and therefore neutrinos are trapped in the stellar medium. When the density decreases neutrinos decouple from the disk matter and can reach a detector or interact with the outflowing ejected matter changing its electron fraction. Therefore, the nucleosynthetic outcome will change as well. There also exists the possibility of probing such an environment on earth by measuring the energy of the neutrino emissions (e.g. in current facilities like SK [14] and SNOLab [15] or future ones such as DUNE [16]), as the interstellar medium is transparent to neutrinos and therefore they would not be affected by the large distance traveled. In order to understand the chance of detection, as well as the setting of the electron fraction in outflowing matter, the neutrino effective flux must be determined. A key factor to this is the neutrino sphere. Such a surface is defined by the locations where the neutrinos are decoupled from the accretion disk. Roughly speaking, the disk can be understood as discriminated in two regions: the *optically thick* region, where neutrinos are trapped inside the neutrino surface, and the *optically thin* region where neutrinos are decoupled from the disk matter. The location of the neutrino surface is not sharp as it depends on the neutrino energy. To account for this fact, we find the neutrino opacity based on cross sections statistically averaged over the neutrino Fermi-Dirac flux,

$$\phi(E_\nu, T) = \frac{g_\nu c}{2\pi^2(\hbar c)^3}\frac{E_\nu^2}{\exp(E_\nu/T)+1}, \tag{1}$$

where $c$ is the speed of light, $\hbar = h/2\pi$, with $h$ the Planck constant, $E_\nu$ is the neutrino energy and the neutrino statisical weight, $g_\nu$, is equal to 1. $T$ is the *local* temperature which, in accretion disks, is usually larger than in supernovae and larger than the neutrino chemical potential. Therefore, is a good approximation to assume this chemical potential equal to zero (see e.g. [17, 18]). The resulting average cross-sections are functions of temperature. The following outlines a process by which the neutrino surface can be determined (see also [19]).

In the high-energy environment of the accretion disk, matter is dissociated into protons, neutrons, electrons, and neutrinos. The neutrino interactions included in our calculation will then consist of all the interactions with these fermions. The charged current reactions are given by

$$\nu_e + n \to p + e^-, \tag{2}$$

$$\overline{\nu_e} + p \to e^+ + n, \tag{3}$$

and the neutral current reactions for all flavours of neutrinos and antineutrinos are given by

$$\nu + p \to \nu + p, \tag{4}$$

$$\nu + n \to \nu + n, \tag{5}$$

$$\nu + e^- \to \nu + e^-, \tag{6}$$

$$\nu + \overline{\nu} \to e^+ + e^-. \tag{7}$$

Assuming charge neutrality, the electron fraction, $Y_e$, is equal to the proton fraction $Y_p$. Therefore, the proton, $n_p$, and neutron, $n_n$, number densities are simply given by

$$n_p = \rho_m N_A Y_e, \tag{8}$$

$$n_n = \rho_m N_A (1 - Y_e), \tag{9}$$

where $\rho_m$ is the mass density, and $N_A$ is Avogadro's number. The electron and positron number densities can be evaluated numerically by integrating their phase space densities with relativistic corrections, and by using $\mu_{e^-} + \mu_{e^+} = 0$ as well as our charge neutrality assumption, $n_p + n_{e^+} = n_e^-$. The opacity is given by

$$\tau_\nu = \int_{h_\nu}^{h_{max}} \sum_k n_k \langle \sigma_k(E_\nu) \rangle dz, \tag{10}$$

where $n_k$ is the number density of the particle with which the neutrino is interacting and $\langle \sigma_k(E_\nu) \rangle$ is the corresponding statistically averaged cross section. The neutrino surface is defined when the opacity $\tau_\nu = 2/3$. Neutrinos within the surface cannot escape due to the large number of interactions, however, those outside can. The neutrino fluxes and average neutrino energies can be calculated from the local temperature at this surface and integrating over the surface area.

The above method was applied to a 2D axis-symmetric hydrodynamical simulation of an accretion disk resulting from a BH-NS merger by O. Just et al [20]. The BH in the simulation has mass $M = 3M_\odot$ and spin parameter $a = 0.8$. Cylindrical coordinates implemented on the system allow for the tracking of density, temperature, and electron fraction at each point in space defined by $(r, \phi, z)$. The simulation is Newtonian, however, the effects of general relativity are introduced via a pseudo-Newtonian potential, specifically, the Artemova potential [21]. The results of our post-proccessing calculation for the opacity $\tau_\nu$ (coloured scale), for electron neutrinos and electron antineutrinos, at a time $t = 20$ ms, are shown in figures 3 and 4. Here we have superimposed the neutrino surface (thick black lines). The region above the neutrino surface is considered to be optically thin to neutrinos, whereas the region below is optically thick.

We see that the electron neutrino surface is much larger than the electron antineutrino surface. This is due to the larger number of scattering processes that electron neutrinos undergo with neutrons (see eqs.2 and 3) compared to the number of interactions that electron antineutrinos have with protons. This is to be expected as the accretion disk mainly consists of neutrons from the neutron star, and thus the charged current reaction for the electron neutrino will happen much more frequently than the corresponding of the electron antineutrino, increasing the size of the neutrino surface.

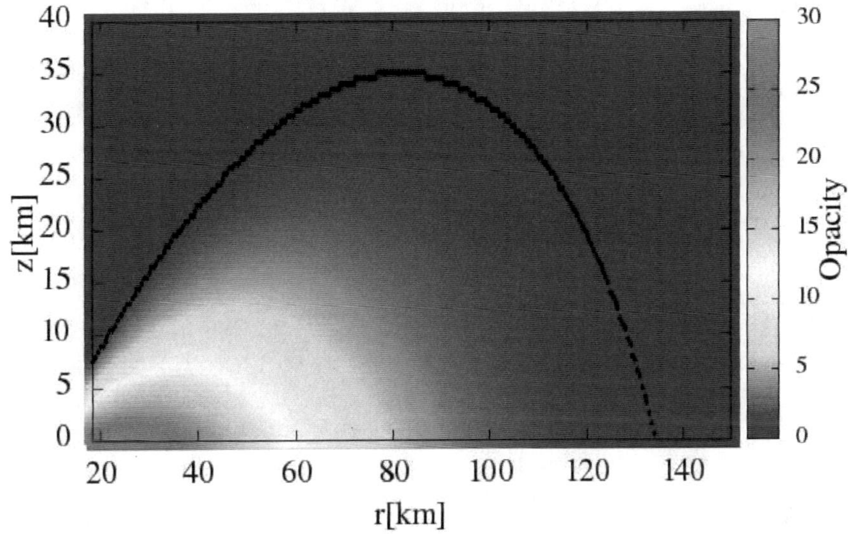


**Figure 3.** Electron neutrino opacity (colour) and surface (black line).

We calculate the innermost stable circular orbit radius, $R_{ISCO}$, in geometrized units for the system in the Kerr metric by [22],

$$R_{ISCO} = M\Big[3 + Z_2 - [(3 - Z_1)(3 + Z_1 + 2Z_2)]^{\frac{1}{2}}\Big], \tag{11}$$

where

$$Z_1 = 1 + \left[1 - \left(\frac{a}{M}\right)^2\right]^{\frac{1}{3}} \left[\left(1 + \frac{a}{M}\right)^{\frac{1}{3}} + \left(1 - \frac{a}{M}\right)^{\frac{1}{3}}\right], \tag{12}$$

$$Z_2 = \left[3\left(\frac{a}{M}\right)^2 + Z_1^2\right]^{\frac{1}{2}}, \tag{13}$$

$a$ is the spin parameter of the BH, and $M$ is its mass. Orbits within this radius may be bound, however, no circular orbit can be stable. As shown in the equation, $R_{ISCO}$ decreases with increasing spin parameter. This is due to the BH dragging spacetime with it as it rotates. Further, we can see the orbit increases with larger BH masses due to larger gravitational forces.

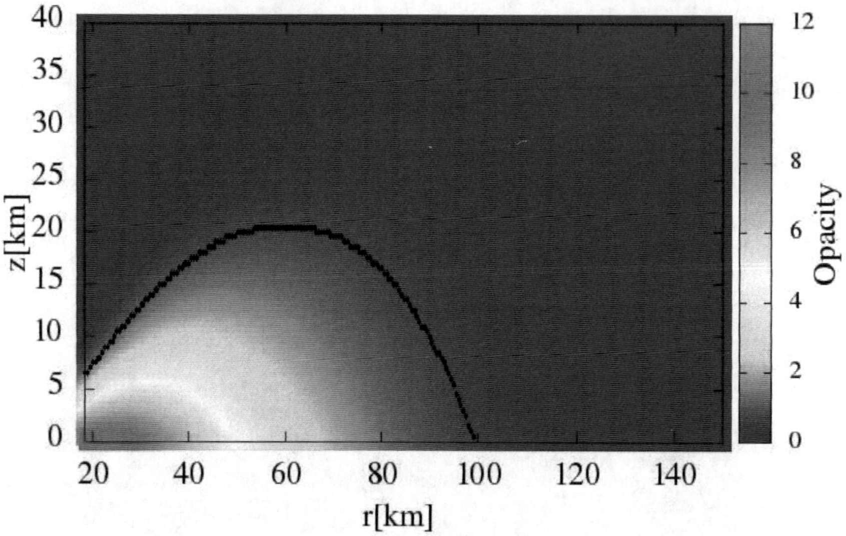


**Figure 4.** Electron antineutrino opacity (colour) and surface (black line).

For the values of $a$ and $M$ from the O. Just et al simulation we find that $R_{ISCO}$=12.8 km. However, the inner edge of the disk, as given in the original simulation, is at 18.5 km. This suggests that matter in the accretion disk between a radii of 12.8 to 18.5 km may emit neutrinos. The fact that the neutrino opacity is low, close to the BH on the symmetry axis, means that neutrinos emitted inside these radii are going to be strongly affected by the BH. At the same time, from the simulation we also know that those neutrinos are the most energetic neutrinos (the local temperature is high: about 10 to 15 MeV), and therefore this could potentially change the properties of the neutrino spectrum.

## 4. Conclusions

We have presented two weak processes that affect the synthesis elements: neutron captures on nuclei and neutrino absorption on nucleons. Neutron captures in neutron degenerate matter is important in the production of elements in accreting neutron stars. We find that the neutron degeneracy changes the rate of absorption by orders of magnitude for $^{46}$Mg and different isotopes of Sn. In future work we will study different nuclei over a broader range of temperatures and densities, as well as delve into more efficient algorithms to account for resonant peaks in the cross sections.

Neutrinos are copiously emitted from black-hole neutron star mergers. They are key in setting the ratio of neutrons to protons in these of environments. They could also be detected if one of these events occurs in a nearby galaxy. We have introduced here results of neutrino opacities and neutrino surfaces for an accretion disk formed during the merger of a black-hole and a neutron star. Our results suggest that further studies of the general relativistic effects are needed to better understand the neutrino spectra close to $R_{ISCO}$ where presumably the most energetic neutrinos are emitted.

**Acknowledgments**
OLC thanks Hendrik Schatz for exciting discussions. We acknowledge support from the Natural Sciences and Engineering Research Council of Canada (NSERC).

## 5. References

[1] B. P. Abbott *et al.* [LIGO Scientific and Virgo Collaborations], Phys. Rev. Lett. **119**, no. 16, 161101 (2017)
[2] B. P. Abbott *et al.* [LIGO Scientific and Virgo and Fermi GBM and INTEGRAL and IceCube and IPN and Insight-Hxmt and ANTARES and Swift and Dark Energy Camera GW-EM and Dark Energy Survey and DLT40 and GRAWITA and Fermi-LAT and ATCA and ASKAP and OzGrav and DWF (Deeper Wider Faster Program) and AST3 and CAASTRO and VINROUGE and MASTER and J-GEM and GROWTH and JAGWAR and CaltechNRAO and TTU-NRAO and NuSTAR and Pan-STARRS and KU and Nordic Optical Telescope and ePESSTO and GROND and Texas Tech University and TOROS and BOOTES and MWA and CALET and IKI-GW Follow-up and H.E.S.S. and LOFAR and LWA and HAWC and Pierre Auger and ALMA and Pi of Sky and DFN and ATLAS Telescopes and High Time Resolution Universe Survey and RIMAS and RATIR and SKA South Africa/MeerKAT Collaborations and AstroSat Cadmium Zinc Telluride Imager Team and AGILE Team and 1M2H Team and Las Cumbres Observatory Group and MAXI Team and TZAC Consortium and SALT Group and Euro VLBI Team and Chandra Team at McGill University], Astrophys. J. **848**, no. 2, L12 (2017)
[3] Q. Z. Liu, J. van Paradijs and E. P. J. v. d. Heuvel, Astron. Astrophys. **469**, 807 (2007)
[4] H. Schatz, L. Bildsten, A. Cumming and M. Wiescher, Astrophys. J. **524**, 1014 (1999)
[5] S. Gupta, E. F. Brown, H. Schatz, P. Moeller and K. L. Kratz, Astrophys. J. **662**, 1188 (2007)
[6] H. Schatz *et al.*, Phys. Rept. **294**, 167 (1998).
[7] R. Mckinven, A. Cumming, Z. Medin and H. Schatz, Astrophys. J. **823**, no. 2, 117 (2016)
[8] P. Haensel, A.Potekhin and D.Yakovlev", "Neutron stars: Equation of state and structure", Springer,Newyork, 2007, pages 517.
[9] C. Arlandini, F. Kaeppeler, K. Wisshak, R. Gallino, M. Lugaro, M. Busso and O. Straniero, Astrophys. J. **525**, 886 (1999)
[10] M. Arnould, S. Goriely and K. Takahashi, Phys. Rept. **450**, 97 (2007)
[11] P. S. Shternin, M. Beard, M. Wiescher and D. G. Yakovlev, Phys. Rev. C **86**, 015808 (2012).
[12] A.J. Koning, S. Hilaire, M.C. Duijvestijn TALYS-1.0 O. Bersillon, F. Gunsing, E. Bauge, R. Jacqmin, S. Leray (Eds.), Proceedings of the International Conference on Nuclear Data for Science and Technology, EDP Sciences, vol. 211, April 2227, 2007, Nice, France (2008)
[13] D. D. Clayton, *Principles of stellar evolution and nucleosynthesis*, McGraw-Hill, New York, 1968, page 288
[14] S. Fukuda et al., Nucl. Inst. Meth. Phys. A, **501** (2003) 418.
[15] F. Descamps [SNO+ Collaboration], Nucl. Part. Phys. Proc. **265-266**, 143 (2015).
[16] R. Acciarri *et al.* [DUNE Collaboration], arXiv:1512.06148 [physics.ins-det].
[17] R.F. Sawyer, Phys. Rev. D **68**, 063001 (2003)
[18] G. M .Fuller, R. Mayler, B. S. Meyer, J. R. Wilson, Astrophys. J., 389, 517 (1992)
[19] O. L. Caballero, G. C. McLaughlin and R. Surman. Phys. Rev. D. 80, 123004 (2009)
[20] O. Just, A. Bauswein, R. Ardevol Pulpillo, S. Goriely, and H.-T. Janka. R. Astron. Soc. 448, 541-567 (2015)
[21] Artemova I. V., Bjoernsson G., Novikov I. D.. 1996. Astrophys.J.,461,565
[22] J. M. Bardeen, W. H. Press and S. A. Teukolsky, Astrophys. J. **178**, 347 (1972).

IOP Conf. Series: Journal of Physics: Conf. Series **1078** (2018) 012013 doi:10.1088/1742-6596/1078/1/012013

# Reaction Dynamics for the Systems $^{7}$Be,$^{8}$B + $^{208}$Pb at Coulomb Barrier Energies

**M Mazzocco[1,2], A Boiano[3], C Boiano[4], M La Commara[3,5], C Manea[2], C Parascandolo[3], D Pierroutsakou[3], C Signorini[6], E Strano[1,2], D Torresi[1,2,25], H Yamaguchi[7], D Kahl[7,26], L Acosta[8,9,27], P Di Meo[3], J P Fernandez-Garcia[9,28], T Glodariu[10,29], J Grebosz[11], A Guglielmetti[4,12], N Imai[7,13], Y Hirayama[13], H Ishiyama[13], N Iwasa[14], S C Jeong[13,15], H M Jia[16], N Keeley[17], Y H Kim[13], S Kimura[13], S Kubono[7,18], J A Lay[1,2], C J Lin[16], G Marquinez-Duran[8], I Martel[8], H Miyatake[13], M Mukai[13], T Nakao[19], M Nicoletto[2], A Pakou[20], K Rusek[21], Y Sakaguchi[7], A M Sánchez-Benítez[8], T Sava[10], O Sgouros[20,25], F Soramel[1,2], V Soukeras[20,25], E Stiliaris[22], L Stroe[10], T Teranishi[23], N Toniolo[6], Y Wakabayashi[18], Y X Watanabe[13], L Yang[16] and Y Y Yang[24]**

[1] Dipartimento di Fisica e Astronomia, Università di Padova, Padova, Italy
[2] INFN-Sezione di Padova, Padova, Italy
[3] INFN-Sezione di Napoli, Napoli, Italy
[4] INFN-Sezione di Milano, Milano, Italy
[5] Dipartimento di Fisica, Università di Napoli "Federico II", Napoli, Italy
[6] INFN-Laboratori Nazionali di Legnaro (LNL), Legnaro (PD), Italy
[7] CNS - The University of Tokyo, RIKEN campus, Wako, Saitama, Japan
[8] Departamento de Física Aplicada, Universidad de Huelva, Huelva, Spain
[9] INFN-Sezione di Catania, Catania, Italy
[10] NIPNE, Magurele, Romania
[11] IFJ PAN, Krakow, Poland
[12] Dipartimento di Fisica, Università di Milano, Milano, Italy
[13] KEK, Tsukuba, Ibaraki, Japan
[14] Department of Physics, Sendai, Miyagi, Japan
[15] Institute for Basic Science, Daejeon, Korea
[16] China Institute of Atomic Energy, Beijing, China
[17] National Centre for Nuclear Research, Otwock, Poland
[18] RIKEN Nishina Center, Wako, Saitama, Japan
[19] Advanced Science Research Center, JAEA, Tokai, Ibaraki, Japan
[20] Department of Physics, University of Ioannina and HINP, Ioannina, Greece
[21] Heavy Ion Laboratory, University of Warsaw, Warsaw, Poland
[22] Department of Physics, University of Athens, Athens, Greece
[23] Department of Physics, Kyushu University, Hakozaki, Fukuoka, Japan
[24] Institute of Modern Physics, Chinese Academy of Sciences, Lanzhou, China

[25] Present address: INFN-Laboratori Nazionali del Sud (LNS), Catania, Italy.
[26] Present address: School of Physics and Astronomy, University of Edinburgh, Edinburgh, UK.
[27] Present address: Instituto de Física, Universidad Nacional Autónoma de Mèxico, Ciudad de Méxixo, Mexico.
[28] Present address: Departamento de Física Atómica, Molecular y Nuclear, Universidad de Sevilla, Sevilla, Spain.
[29] Deceased.

IOP Conf. Series: Journal of Physics: Conf. Series **1078** (2018) 012013 doi:10.1088/1742-6596/1078/1/012013

E-mail: marco.mazzocco@pd.infn.it

**Abstract.** In this contribution we describe the first results obtained for the investigation of the elastic scattering process in the reactions induced by the Radioactive Ion Beams $^{7}$Be and $^{8}$B on a $^{208}$Pb target at Coulomb barrier energies. The experimental data were analyzed within the framework of the optical model in order to extract the total reaction cross section. The comparison with data available in literature for reactions induced on $^{208}$Pb by light ions in the mass range A = 6-8 shows that the loosely-bound $^{8}$B has the largest reactivity.

## 1. Introduction

The study of the reaction dynamics induced by light weakly-bound projectiles in the energy range around the Coulomb barrier has attracted the interest of the Nuclear Physics community since the early Nineties and the investigation was later boosted by growing availability of Radioactive Ion Beams (RIBs). Several review papers have been written on this topic in the past years [1, 2, 3, 4, 5, 6, 7].

These studies were originally aimed at measuring the sub-barrier fusion process. In fact, even in reactions induced by stable projectiles, a relevant enhancement of the fusion probability was observed at energies below the Coulomb barrier [8]. A detailed analysis showed that both static features, such as the nuclear deformation, and dynamical properties, such as the presence of transfer channels with positive $Q_{value}$, could dramatically alter the fusion cross section. With the advent of RIBs, it was expected that these effects could even be magnified by the unusual characteristics of these nuclei. In fact, especially in the lightest corner of nuclide chart we can find several examples of exotic nuclei with strong deformations or well established cluster configurations. Moreover, as a rather general feature, exotic projectiles are in most cases very loosely bound, with (total) binding energies often smaller than 1 MeV. It is therefore quite likely that, while approaching a target nucleus, a weakly bound projectile could more easily be broken by the interaction with the Coulomb and nuclear field provided by the reaction partner. We thus expect to observe large breakup cross sections, which should influence the whole reaction dynamics. From a theoretical point of view, if the breakup process is considered to be an additional open channel, we should quantum mechanically measure an enhancement of the sub-barrier fusion cross section. On the other side, if we treat the projectile breakup as a flux removal from the reaction input channel, then we might have a hindrance of the fusion probability, because we would have fewer particles at disposal for all other mechanisms.

Earlier measurements indicated a rather large increase of the sub-barrier fusion cross section [9], however it was soon realized that breakup related effects mostly enhanced the reaction rather than the fusion probability. This enhancement was mainly due to the 2n-stripping process, as for instance in $^{6}$He- [10, 11, 12] and $^{8}$He-induced [13, 14] reactions, whereas there are strong signatures that the breakup process could be responsible for the enhancement in reactions induced by the neutron-halo projectiles $^{11}$Li [15, 16] and $^{11}$Be [18, 17]. In all cases, serious deviations from the usual Rutherford scattering differential cross section were observed even at deep sub-barrier energies and this is another important evidence of strong reaction couplings.

Within this framework, we started the investigation of $^{7}$Be- and $^{8}$B-induced reactions. $^{8}$B is an extremely weakly-bound proton-halo nucleus ($S_p = 0.1375$ MeV), while $^{7}$Be is the core of $^{8}$B and is a loosely-bound nucleus ($S_\alpha$ 1.586 MeV) as well. As a difference with most of the RIBs studied so far, these two nuclei are located on proton rich side of the valley of $\beta$-stability and they might exhibit different features with respect to neutron halo and neutron skin nuclei. Until now, only the interaction of $^{7}$Be and $^{8}$B with light ($^{28}$Si [19, 20, 21]) and medium-mass ($^{58}$Ni [22, 23, 24]) has been investigated. Our study represents the first measurement performed at Coulomb barrier energies on the heavy target $^{208}$Pb.

## 2. Experiments

The $^7$Be RIB for this experiment was produced by using the EXOTIC facility [25, 26, 27] located at the Laboratori Nazionali di Legnaro (LNL) of the Istituto Nazionale di Fisica (INFN) in Legnaro, close to Padova (Italy). The experiment with the $^8$B was performed in Japan by means of the CNS Radioactive Ion Beam facility (CRIB) [28, 29] of the Center for Nuclear Studies (CNS) of The University of Tokyo and located inside the RIKEN campus in Wako-shi. In the next paragraphs we summarize the most important details concerning the RIB production and the experimental set-up employed for the two experiments.

### *2.1. $^7$Be Experiment*

*2.1.1. Beam Production* The $^7$Be delivered by the facility EXOTIC was produced employing the two-body reaction p($^7$Li,$^7$Be)n, where a 48.8 MeV $^7$Li primary beam, accelerated by the LNL-XTU Tandem, was impinging on a gas target filled with 1 bar of hydrogen. The outcoming RIB was separated from the scattered beam and from the other contaminants up to a degree of purity of about 99% by using the eight ion optical elements of the facility. Details on the RIB production at EXOTIC can be found in the most recent publications [30, 31]. The secondary beam was impinging on a 1 mg/cm$^2$ thick $^{208}$Pb target foil with an intensity of 2-3 $\times$ $10^5$ pps. Three beam energies were obtained by operating the target station in different conditions (liquid nitrogen or room temperature) and inserting a aluminum degrader at a suitable position along the beam line. The beam energies at mid-target positions were: 37.4, 40.5 and 42.2 MeV.

*2.1.2. Experimental Set-up* Charge reaction products were detected with 6 modules of the detector array EXPADES [32, 33]. Each module consisted of a $\Delta$E-E$_{res}$ telescope of Double Sided Silicon Strip Detectors (DSSSDs). The thickness of the telescope inner ($\Delta$E) and outer (E$_{res}$) stage was 43-57 and 300$\mu$m, respectively. The active area of each DSSSD was 64 mm $\times$ 64 mm and was divided into 32 strips per side, thus allowing a 2 mm $\times$ 2 mm pixel resolution. The telescopes were place symmetrically around the beam axis at the following mean polar angles: $\theta_{lab} = \pm 69°, \pm 111°$ and $\pm 153°$ in order to cover approximately the angular $\theta_{lab} = [50°, 170°]$.

### *2.2. $^8$Be Experiment*

*2.2.1. Beam Production* The $^8$B RIB for the experiment performed in Japan was produced by employing the inverse kinematics reaction $^3$He($^6$Li,$^8$B)n, where the $^6$Li primary beam was delivered with an energy of 11 MeV/u by the RIKEN AVF Cyclotron and was hitting an 8-cm long gas cell inflated with $^3$He gas kept at liquid nitrogen temperature and at a pressure of about 1 bar. After the selection with the CRIB separator, the $^8$B beam was hitting a 2.2 mg/cm$^2$ thick $^{208}$Pb target with an intensity of $10^4$ pps and a purity of about 20%. The secondary beam energy on target was about 50 MeV.

*2.2.2. Experimental Set-up* Also in this case, we employed six modules of the detector array EXPADES. However, in order to ensure a wider polar angle coverage, a slightly asymmetric arrangement was used. The mean polar angles of the six telescopes, in fact, were: $\theta_{lab} = +27°, \pm 69°, \pm 111°$ and $-153°$, where the positive and negative signs preceding the polar angles indicate the left and right hemisphere (in a downstream view) of the scattering chamber. In such a way, it was possible to cover the angular range: $\theta_{lab}$ = 15°-165°.

## 3. Preliminary Results

### *3.1. Elastic Scattering*

Figures 1 and 2 show the preliminary evaluation of the elastic scattering angular distributions for the two reactions. Due to the secondary beam energy resolution and to the target thickness, it

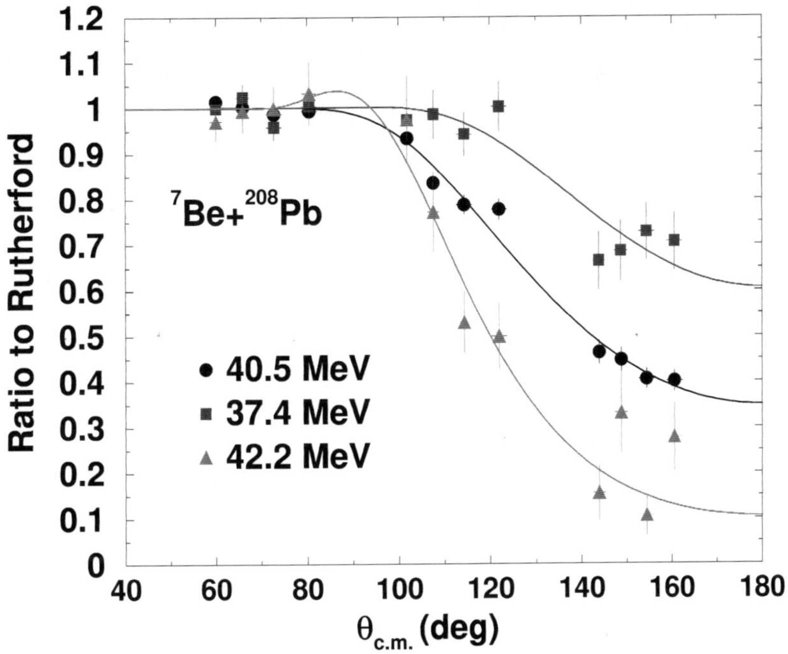


**Figure 1.** (Quasi-)elastic differential cross sections for the reaction $^{7}$Be + $^{208}$Pb at 37.4, 40.5 and 42.2 MeV. Continuous lines are optical model best-fits of the experimental data.

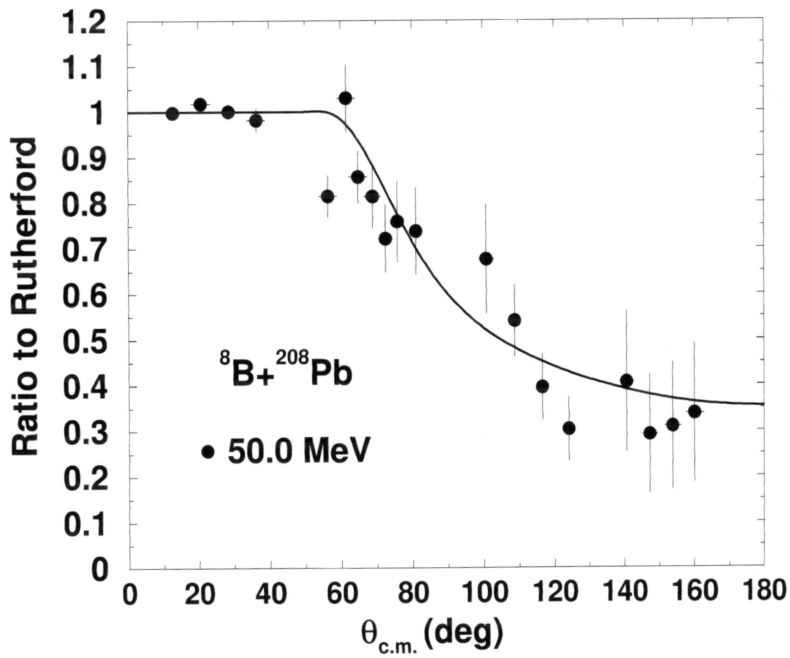


**Figure 2.** Elastic scattering differential cross section for the reaction $^{8}$B + $^{208}$Pb at 50 MeV. The continuous line is the result of the optical model best-fit analysis of the experimental data.

was not possible to solve experimentally pure elastic scattering events from inelastic excitations leading to the $^7$Be first excited state at $E_x = 0.429$ MeV. For this reason, the experimental data displayed in Figure 1 have to be considered as quasi-elastic. To compensate the limited statistics gathered, the elastic scattering data were grouped into bins of four adjacent vertical strips and only four points per detector are represented in Figures 1 and 2. Where available, the average of the differential cross sections evaluated by telescopes located in the left and in the right hemisphere of the scattering chamber was considered.

The effects of the nuclear interaction between the projectiles and the target are clearly visible in both figures. The differential cross sections decrease as the scattering angles increases, since obviously the distance of closest approach decreases, and as the beam energy increases, as new reaction channels open up. The experimental data were fitted within the framework of the optical model in order to extract the total reaction cross section. The starting point of the fitting procedure was the Broglia-Winther parametrization [34] of the nuclear potential. The radius and diffuseness of both the real part and the imaginary part of the potential well were kept fixed and only the depths were let free to vary. The coupled channel code FRESCO [35] was employed for these calculations. Continuous lines in Figures 1 and 2 are the present results of the fitting procedure.

### *3.2. Reaction Cross Section*

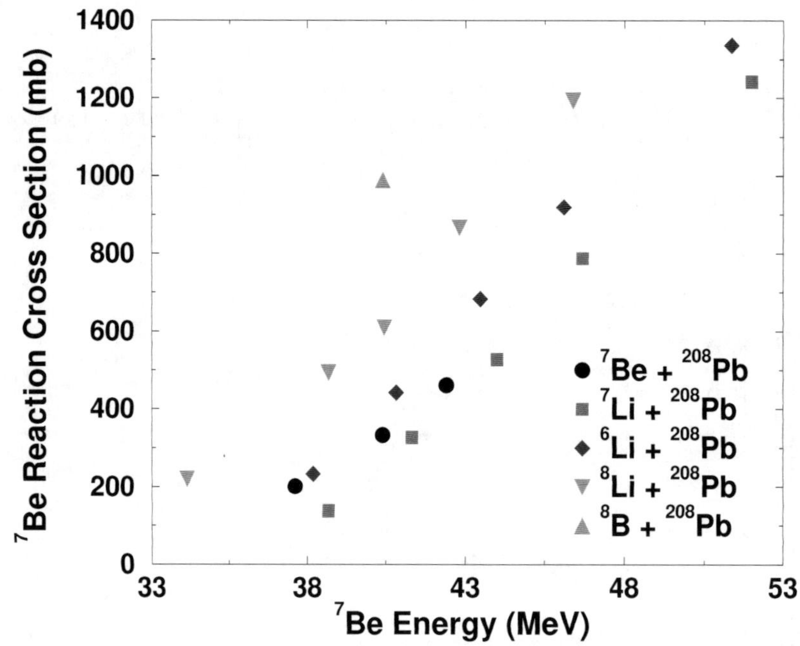


**Figure 3.** Preliminary evaluation of the total reaction cross sections for the systems $^{6,7,8}$Li, $^7$Be and $^8$B + $^{208}$Pb at Coulomb barrier energies. To compare systems with different Coulomb barrier and geometrical size, energies and cross sections were normalized to the values for the system $^7$Be + $^{208}$Pb.

Figure 3 shows the comparison of the preliminary evaluation of the total reaction cross section for the three beam energies we measured for the system $^7$Be + $^{208}$Pb and the only energy point collected for the reaction $^8$B + $^{208}$Pb. For the sake of comparison, the total reaction cross section data available in literature for the systems $^{6,7,8}$Li + $^{208}$Pb [36, 37] were also included in the figure. To account for the different Coulomb barrier and the different projectile size, the beam energies and the cross sections were normalized following the procedure described in [38] and then multiplied by the scaling factors for the system $^{7}$Be + $^{208}$Pb.

We can immediately appreciate that $^{7}$Be exhibits an intermediate behavior between that for $^{6}$Li, nucleus characterized by a very similar binding energy, and that for its mirror nucleus $^{7}$Li, which has a rather similar nuclear structure but is more bound by nearly 1 MeV. At near-barrier energies, where the projectile binding energy plays a major role, the total reaction cross section for the system $^{7}$Be + $^{208}$Pb resembles that for $^{6}$Li, whereas, as the beam energy increases, nuclear structure effects become more relevant and the $^{7}$Be reactivity approaches that for $^{7}$Li.

The comparison between the mirror nuclei $^{8}$Li and $^{8}$B is also very striking, being the reaction cross section for the boron isotope nearly twice that for the lithium counterpart. Such a large difference is presently interpreted as a direct consequence of the very small $^{8}$B binding energy and a possible signature of a strong breakup channel $^{8}\text{B} \rightarrow {}^{7}\text{Be} + p$.

## 4. Perspectives

First-hand results on the reaction dynamics induced at Coulomb barrier energy by the two RIBs $^{7}$Be and $^{8}$B on a $^{208}$Pb target have been presented. The data analysis concentrated so far on the evaluation of the elastic scattering angular distributions in order to extract the total reaction cross sections. A preliminary comparison for the reactions induced at Coulomb barrier energies on a $^{208}$Pb target by light weakly-bound projectiles in the mass range A = 6-8 has been presented.

The next steps of the analysis will consist in finalizing the theoretical calculations, by investigating, for instance, the relevance of inelastic excitations in the reaction induced $^{7}$Be and in producing reliable estimates for the breakup cross section in the system $^{8}$B + $^{208}$Pb. For an experimental point of view, the data analysis of the angular distributions for charged reaction products other than the projectile could provide additional insights on the reaction dynamics induced by these very exotic projectiles.

## Acknowledgments

We all mourn our dear mate Tudor Glodariu, who unfortunately passed away on November $14^{th}$, 2017 at the age of 46. Tudor had been deeply involved in the lay-out and commissioning of the facility EXOTIC at INFN-LNL since the beginning of the project. He participated in all experiments performed, being in several cases the spokesperson or the co-spokesperson of the measurement. We will always remember his kind personality, his dedication to work, his fine sense of humor and the pleasure to have met in our lives a very special person like Tudor.

## References

[1] Canto L F, Gomes P R S, Donangelo R and Hussein M S 2006 *Phys. Rep.* **424** 1
[2] Liang J F and Signorini c 2005 *Int. J. Mod. Phys. E* **14** 1121
[3] Keeley N, Raabe R, Alamanos N and Sida J L 2007 *Prog. Part. Nucl. Phys.* **59** 579
[4] Keeley N, Alamanos N, Kemper K W and Rusek K 2009 *Prog. Part. Nucl. Phys.* **63** 396
[5] Mazzocco M 2010 *Int. J. Mod. Phys. E* **19** 977
[6] Keeley N, Kemper K W and Rusek K 2014 *Eur. Phys. J. A* **50** 145
[7] Canto L F, Gomes P R S, Donangelo R, Lubian J and Hussein M S 2015 *Phys. Rep.* **596** 1
[8] Dasgupta M, Hinde D J, Rowley N and Stefanini A M 1998 *Ann. Rev. Nucl. Part. Sci.* **48** 401
[9] Trotta M *et al* 2000 *Phys. Rev. Lett.* **84** 2342
[10] A. Di Pietro A *et al* 2004 *Phys. Rev. C* **69** 044613
[11] Navin A *et al* 2004 *Phys. Rev. C* **70** 044601
[12] Raabe R *et al* 2004 *Nature* **431** 823
[13] Lemasson A *et al* 2009 *Phys. Rev. Lett.* **103** 232701
[14] Lemasson A *et al* 2010 *Phys. Rev. C* **82** 044617

[15] Cubero M *et al* 2012 *Phys. Rev. Lett.* **109** 262701
[16] Fernandez-Garcia J P *et al* 2013 *Phys. Rev. Lett.* **110** 142701
[17] Pesudo V *et al* 2017 *Phys. Rev. Lett.* **118** 152502
[18] Di Pietro A *et al* 2010 *Phys. Rev. Lett.* **105** 022701
[19] Pakou A *et al* 2013 *Phys. Rev. C* **87** 014619
[20] Sgouros O *et al* 2016 *Phys. Rev. C* **94** 044623
[21] Sgouros O *et al* 2017 *Phys. Rev. C* **95** 054609
[22] Aguilera E F *et al* 2009 *Phys. Rev. C* **79** 021601(R)
[23] Aguilera E F *et al* 2011 *Phys. Rev. Lett.* **107** 092701
[24] Mazzocco M *et al* 2015 *Phys. Rev. C* **92** 024615
[25] Maidikov V Z *et al* 2004 *Nucl. Phys. A* **746** 389c
[26] Pierroutsakou D *et al* 2007 *Eur. Phys. J. Spec. Top.* **150** 47
[27] Farinon F *et al* 2008 *Nucl. Instrum. Meth. B* **266** 4097
[28] Yanagisawa Y *et al* 2005 *Nucl. Instrum. Meth. A* **539** 74
[29] Yamaguchi H *et al* 2008 *Nucl. Instrum. Meth. A* **589** 150
[30] Mazzocco M *et al* 2008 *Nucl. Instrum. Meth. B* **266** 4665
[31] Mazzocco M *et al* 2013 *Nucl. Instrum. Meth. B* **317** 223
[32] Strano E *et al* 2013 *Nucl. Instrum. Meth. B* **317** 657
[33] Pierroutsakou D *et al* 2016 *Nucl. Instrum. Meth. A* **834** 46
[34] Broglia R A and Winther A 1981 *Heavy Ion Reactions* (Reading, MA: Benjamin)
[35] Thompson I J 1988 *Comput. Phys. Rep.* **2** 167
[36] Keeley N *et al.* 1994 *Nucl. Phys. A* **571** 32
[37] Kolata J J *et al* 2002 *Phys. Rev. C* **65** 054616
[38] Gomes P R S, Lubian J, Padron I, Anjos R M 2005 *Phys. Rev. C* **71** 017601

IOP Conf. Series: Journal of Physics: Conf. Series **1078** (2018) 012014 doi:10.1088/1742-6596/1078/1/012014

# Fusion measurements for the $^7$Li+$^{51}$V system at energies around the Coulomb barrier

**J C Morales-Rivera[1,2], E Martinez-Quiroz[1], P Amador-Valenzuela[1], E F Aguilera[1], T L Belyaeva[2] and D Lizcano[1]**

[1] Departamento de Aceleradores, Instituto Nacional de Investigaciones Nucleares, Apartado Postal 18-1027, C.P. 11801, México

[2] Facultad de Ciencias, Universidad Autónoma del Estado de México, Instituto Literario No. 100, C.P. 50000, México

E-mail: enrique.martinez@inin.gob.mx

**Abstract.** We present the experimental fusion cross sections for the $^7$Li+$^{51}$V system measured using the EN Tandem Van de Graaff Accelerator facility at National Institute of Nuclear Research (ININ), in Mexico. The study was focused on energies around the Coulomb barrier, $B$ = 10.33 MeV ($B_{\text{lab}}$ = 11.75 MeV), in the interval from 10.0 MeV to 13.5 MeV in steps of 0.5 MeV, in the laboratory frame. In order to complement this study, the experimental results were compared with a fusion-evaporation analysis using the PACE2, LILITA and CASCADE codes, in the same interval of energies. We found that the $xn$ and $xn\alpha$ are the most relevant evaporation channels, leading to $^{55}$Fe, $^{56}$Fe, $^{52}$Cr and $^{53}$Cr as residual nuclei.

## 1. Introduction

In recent years, the Heavy Ion Group of the National Institute of Nuclear Research (ININ), has incorporated into its research interest the study of nuclear reactions using stable beams on different targets using the EN Tandem Van de Graaff Accelerator facility at ININ, in Mexico. The main goal of these studies is to compare and complement the information on the behavior and nuclear structure of radioactive beams, such as $^7$Be and $^8$B. For this, the elastic scattering angular distributions of the $^9$Be+$^{51}$V and $^7$Li+$^{58}$Ni systems have already been measured around the respective Coulomb barrier energies [1, 2]. Recently, several theoretical and experimental analyzes have been carried out to study the fusion mechanisms of $^7$Li on different targets by various research groups around the world [3, 4, 5, 6, 7, 8]. Following this reasoning we aimed to measure, by using the $\gamma$-ray technique [4, 9, 10], the fusion cross sections for the $^7$Li+$^{51}$V system at energies between $E_{\text{lab}}$ = 10.0 and 13.5 MeV.

In order to compare our experimental data, we performed an evaporation-fusion analysis by using the PACE2 [11], LILITA [12] and CASCADE [13, 14] codes. Calculations made with these codes usually involve the selection of multiple sets of input parameters which may have a wide range of reasonable variation, sometimes leading to important sensitivity in the final results. Examples of the latter are the level density and optical parameters. For this reason, it is necessary to choose an appropriate set of input parameters to assess the compatibility of the corresponding physical calculations that permits a reliable comparison of the fusion cross section deduced for the different codes.

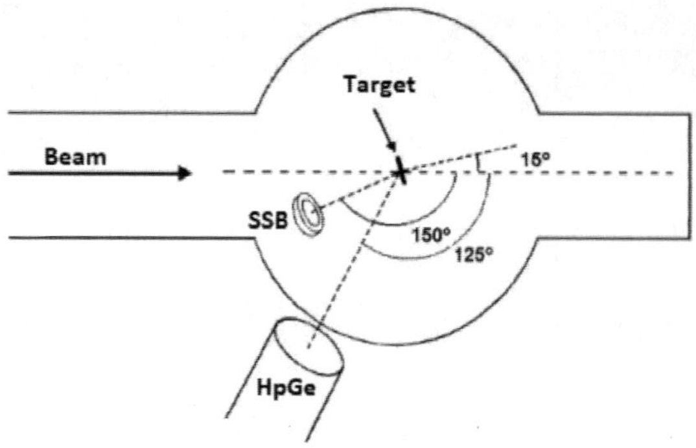


**Figure 1.** Experimental arrangement used in the present experiment.

## 2. Experimental setup

The experiment was carried out at the Accelerator facility at ININ, in Mexico. The $^{7}$Li beam was produced with the 6MV EN Tandem Van de Graaff Accelerator. The experimental arrangement consisted of a natural composition Vanadium thick target, a Surface Silicon-Barrier detector (SSB) and a Hiperpure Germanium detector (HpGe). The SSB detector was used to measure the back-scattered charged particles, while the HpGe detector was used to measure the $\gamma$-rays produced by the residual nuclei formed during the reaction. These detectors were placed at 150 and 125 degrees with respect to the beam axis, as shown in figure 1. The energies considered in this study were $E_{\rm lab}$= 10.0, 10.5, 11.0, 11.5, 11.75, 12.0, 12.5, 13.0 and 13.5 MeV.

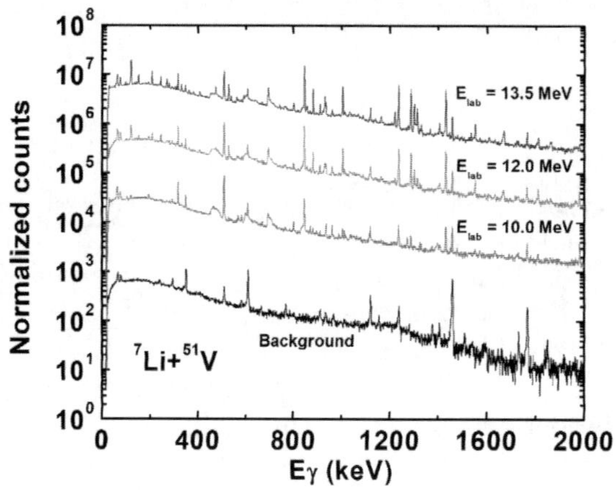


**Figure 2.** Representative $\gamma$-ray spectra, measured at indicated energies. The background spectrum is included for qualitative comparison.

## 3. Results and discussion

The selected $\gamma$-ray spectra measured are shown in figure 2 for indicated energies. Also, background spectrum is shown for comparison. To determine the fusion cross section, only the $\gamma$-rays produced by decays to ground state, were considered.

The experimental fusion cross sections, $\sigma_{fus}$, were determined by the relationship:

$$N_\gamma = \epsilon N_p \int_0^{E_0} \frac{\sigma_{fus} dE}{S(E)}, \tag{1}$$

where $N_\gamma$ is the number of $\gamma$-ray detected (for a specific energy), $\epsilon$ is the absolute efficiency of the detection system, $N_p$ is the number of incident projectiles and $S(E)$ is the stopping power in the target.

The comparison of the data (symbols) with our calculation using the PACE2 code is shown in figure 3, for most important residual nuclei. In most cases, the calculations show some significant differences while in others there are not. In particular, the $^{52}$Cr($2n\alpha$) and $^{53}$V($p\alpha$) isotopes productions turned out to be larger than those predicted by the code. We suppose that there might be some other processes present, such as transfer, breakup and/or incomplete fusion, that are not considered by the code. It is necessary to make a wide analysis by using the CDCC and CRC methods. The same figure shows the experimental total fusion excitation function measured in this study (black circles). It turned out that the $xn$ and $xn\alpha$ are the most relevant evaporation channels leading to residual nuclei $^{55}$Fe, $^{56}$Fe, $^{52}$Cr and $^{53}$Cr, which represent approximately 85% of the total fusion cross section. The points shown here only include the statistical errors, therefore, the error bars are not visible in the experimental data.

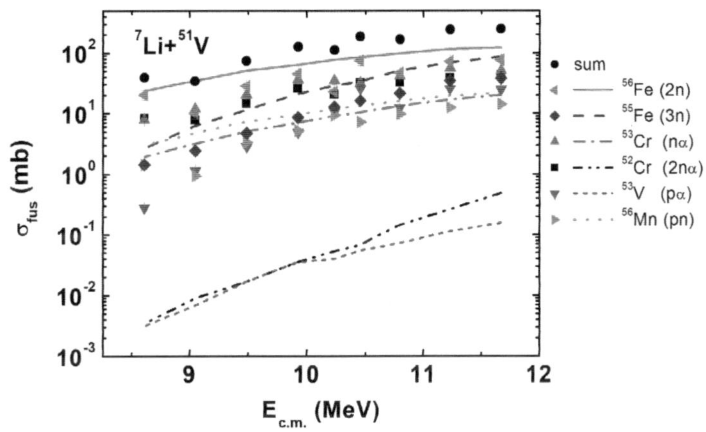


**Figure 3.** (color online) Comparison of the fusion cross sections measured for every residual nuclei with the PACE2 code for the $^7$Li+$^{51}$V system. The black circles represent the sum of the partial cross sections.

On the other hand, in figures 4(a,b), the fusion cross sections predicted by the PACE2 (solid lines), LILITA (dotted lines) and CASCADE (dash-dot lines) codes are shown for the most significant residual nuclei ($^{55}$Fe, $^{56}$Fe, $^{52}$Cr and $^{53}$Cr). One reason of the differences is the way each code does the calculation. According to the three codes, the evaporation of neutrons is the most important process through which the composite nucleus ($^{58}$Fe) releases its exceeding energy. Therefore, the most expected residual nuclei to be produced are $^{56}$Fe($2n$) and $^{55}$Fe($3n$).

We use a well-known data reduction procedure in order to make direct comparisons with similar systems to our study, as proposed in Ref. [15]. For this, we used the data previously reported for $^7$Li+$^{59}$Co [5] and $^7$Li+$^{64}$Zn [8] systems, which have the same projectile. This comparison is shown in figure 5, where our experimental data seem to be located between the two regions formed by the data of these systems.

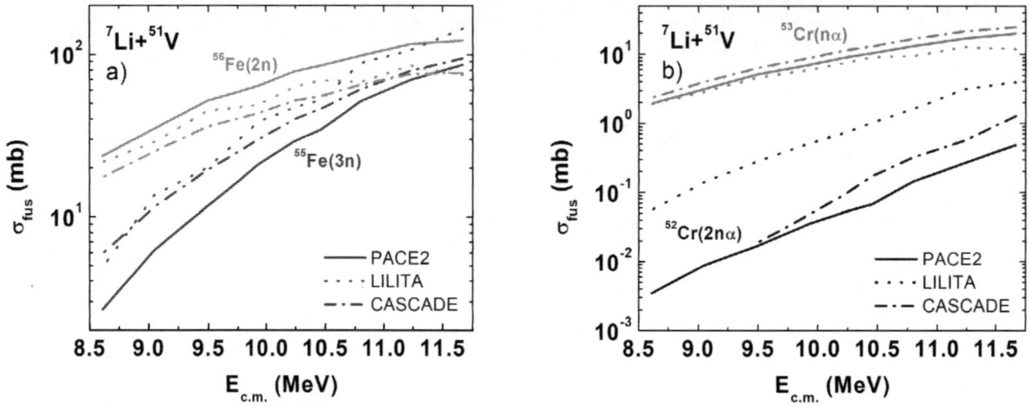


**Figure 4.** (color online)(a)-(b) Comparison of the fusion cross sections of the residual nuclei predicted by PACE2 (solid line), LILITA (dotted line) and CASCADE (dash-dot line) codes.

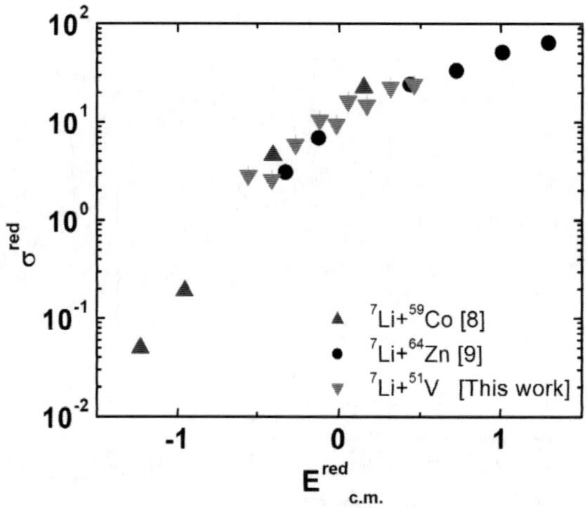


**Figure 5.** (color online) Comparison of the fusion cross sections for the $^{7}$Li+$^{51}$V, $^{7}$Li+$^{59}$Co [5] and $^{7}$Li+$^{64}$Zn [8] systems, in reduced units. A data reduction procedure, proposed in Ref. [15], was used to make a direct comparison between different systems.

## 4. Summary

We performed an experiment to measure the fusion at several energies around the Coulomb barrier for the $^{7}$Li+$^{51}$V system. A theoretical analysis to compare the predictions for the residual nuclei produced in the fusion of the system, by using the PACE2, LILITA and CASCADE codes, was made. The obtained results show a larger production of the $^{52}$Cr($2n\alpha$) and $^{53}$V($p\alpha$) residual nuclei than those predicted by the fusion-evaporation codes. We suppose that these differences may result from transfer, breakup and/or incomplete fusion processes. It is necessary to perform the corresponding calculations to confirm or refute such hypothesis. It turned out that the $xn$ and $xn\alpha$ are the most relevant evaporation channels leading to residual nuclei $^{55}$Fe, $^{56}$Fe, $^{52}$Cr and $^{53}$Cr.

## Acknowledgments
This work was partially supported by CONACYT, Mexico.

## References
[1] Morales-Rivera J C, Martinez-Quiroz E, Belyaeva T L, Aguilera E F, Lizcano D and Amador-Valenzuela P 2016 *Eur. Phys. J. Web Conf.* **117** 07027
[2] Amador-Valenzuela P, Aguilera E F, Martinez-Quiroz E, Lizcano D and Morales-Rivera J C 2017 *J. Phys.: Conf. Ser.* **876** 012002
[3] Diaz-Torres A, Thompson I J and Beck C 2003 *Phys. Rev. C* **68** 044607
[4] Canto L F, Gomes P R S, Donangelo R and Hussein M S 2006 *Physics Reports* **424** 1
[5] Beck C *et. al.* 2003 *Phys. Rev. C* **67** 054602
[6] Beck C 2007 *Nucl. Phys. A* **787** 251
[7] Pakou A *et. al.* 2009 *Eur. Phys. J. A* **39** 187
[8] Di Pietro *et. al.* 2013 *Phys. Rev. C* **87** 064614
[9] Gomes P R S, Penna T J P, Liguori Neto R, Acquadro J C, Tenreiro C, Freitas P A B, Crema E, Carlin Filho N and Coimbra M M 1989 *Nucl. Instr. Method Phys. Res. A* **280** 395
[10] Aguilera E F *et. al.* 2006 *Phys. Rev. C* **73** 064601
[11] Gavron A 1980 *Phys. Rev. C* **21**
[12] Gomez del Campo J and Stokstad R G 1981 *Description and use of the Monte Carlo code LILITA* ORNL-TM-7295 (Oak Ridge, Tennessee, USA)
[13] Pühlhofer F 1977 *Nucl. Phys. A* **280** 267
[14] Harakeh M N, Dowell D H, Feldman G, Garman E F, Loveman R, Osborne J L and Snover K A 1986 *Phys. Lett. B* **176** 297
[15] Aguilera E F, Amador-Valenzuela P, Martinez-Quiroz E, Fernandez-Arnaiz J, Kolata J J and Guimarães V 2016 *Phys. Rev. C* **93** 034613

IOP Conf. Series: Journal of Physics: Conf. Series **1078** (2018) 012015 doi:10.1088/1742-6596/1078/1/012015

# Advances in the measurement of the $^{13}$C(d , p)$^{14}$C cross section using AMS

**S. Murillo-Morales[1], L. Barrón-Palos[1], E. Chávez[1], E. Lopez-Saavedra[1], D. J. Marin-Lambarri[1], L. Acosta[1], G. Murillo[2], R. Policroniades[2], A. Huerta[1], C. Solis[1], A. Varela[2]**

[1] Instituto de Física, Universidad Nacional Autónoma de México. 04510, Ciudad de México.
[2] Instituto Nacional de Investigaciones Nucleares. Carretera México-Toluca s/n, La Marquesa, Ocoyoacac, Estado de México, C.P. 52750.

E-mail: silviamur@ciencias.unam.mx

**Abstract.** An experimental protocol to study the total cross section for the $^{13}$C(d , p)$^{14}$C nuclear reaction via AMS is being currently developed for deuteron energies between 1 and 4 MeV. We started a series of experiments in which two aluminium cathodes filled with natural-graphite (98.9% 12C, 1.1% 13C) samples were irradiated with a deuterium beam of 4 MeV in a 150° RBS configuration at Instituto Nacional de Investigaciones Nucleares 6.0 Tandem Van de Graaff Accelerator (ININ, Mexico). The relative concentrations of $^{14}$C/$^{12}$C were subsequently analyzed using AMS at Laboratorio Nacional de Espectrometría de Masas con Aceleradores (LEMA, UNAM).

## 1. Introduction

The use of the Accelerator Mass Spectrometry (AMS) facilities to measure nuclear reaction cross sections is still pretty new, it was first implemented in 1980 by M. Paul to study $^{26}$Mg$(p,n)^{26}$Al at energies between 5.2 and 6.9 MeV at ANL [1]. This was also the first work that used this technique for laboratory experiments in nuclear astrophysics.

In the present work, an experimental methodology to measure the cross section of $^{13}$C$(d,p)^{14}$C by obtaining the deconvolution of the thick target reaction yield at 4 MeV lab energy using AMS is proposed. Moreover, this is one of the first works using this technique in both nuclear physics and nuclear astrophysics [2] in the only AMS facility available in Mexico, the Accelerator Mass Spectrometry National Laboratory (LEMA) [3].

The aim of this experiment is to be able to compare with existing experimental data [4] [5], stablishing a metodology to perform this kind of studies. Such metodology could be applied in the future to other systems were the cross section is not well known. On the other hand, the $^{13}$C$(d,p)^{14}$C reaction is of interest to study compound nucleus effects [4] and also as a new possible channel for studying the CNO synthesis in the Big Bang Nucleosynthesis (BBN) [6]. Experimental data on this reaction are scarce, so confirming previous result is also of scientific interest.

## 2. Experiment

Originally, we had planned to irradiate graphite samples for deuterium beam energies between 1 and 3 MeV at the 5.5 MV Van de Graaff accelerator at IFUNAM, as stated in our previous paper [7], however, this was not possible due to a malfunctioning of the accelerator's ion source. We had to redraw the experiment for the working conditions of the ININ 6.0 MeV Tandem Van de Graaff Accelerator, where stable beam currents could only be achieved at higher energies. Thus, we irradiated two samples at $E_d = 4$ MeV for 3 hours each. One solid-state detector was placed at 150° with respect to the beam direction in order to detect the backscattered deuterons. The relative concentrations had to be sufficient to obtain minimum measurable $^{14}$C/$^{12}$C suitable for the LEMA AMS-facility sensitivity-range conditions ($10^{-16}$ - $10^{-11}$), and consequently to produce optimum quantities of $^{14}$C. The irradiation time was previously estimated by computing the reaction rates for the relative concentrations of $^{14}$C/$^{12}$C (hereinafter $N_{14}/N_{12}$) for energies between 1 and 3 MeV, where concentrations of the order of $10^{-12}$ were achieved for all these energies after 3 hours.

The targets consisted of two cylindrical aluminium cathodes, each containing 4.6 g of graphite in its natural concentration (98.9% $^{12}$C, 1.1% $^{13}$C) . The surface area of the graphite was of 1.3 mm and the thickness was of 1600 $\mu$m. The geometry of the cathodes is shown in Figure 2.

The experimental setup we used at ININ bears a close resemblance to the one at IFUNAM. The difference lies mainly on the reaction chamber at ININ, which has a mechanical system to align the targets from outside the chamber in order to irradiate several samples one after another without breaking vacuum. Hence we used the sample holder shown in Figure 1 (right) made out of nylon.

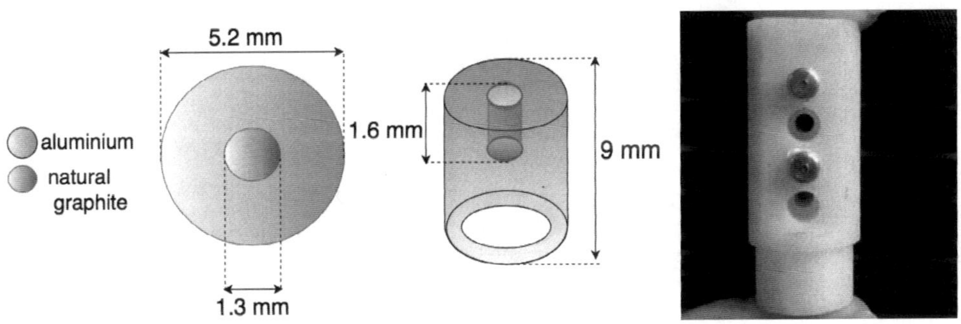


**Figure 1.** Diagram of a graphite-filled aluminium cathode (left); Sample holder (right)

After the irradiation, the samples as well as another two non-irradiated ones were analyzed by AMS at the LEMA facility.

## 3. Results and discussion

The number of deuterons impinging on each sample was determined from the backscattered particle spectra using SIMNRA [8] (Table 1) to simulate each spectrum using experimental data of the cross sections of the reaction channels available at IBANDL [9] at the same conditions, namely data that were available at 4 MeV at the and at 150° as well as the elemental concentration of the composition of the target, which showed that the carbon concentrations were of 31% and 32%, for samples C2 and C3 respectively. The rest corresponds to the aluminium from the cathodes, and also, oxygen and nitrogen as contaminants.

Figure 2 shows an RBS spectrum obtained and set in logarithmic scale, where the reaction and elastic channels identified in the target can be seen as well as some elastic channels from other nuclei that were also present. The reaction channels for $^{12}$C+$d$ and $^{13}$C+$d$ were clearly located. For instance, $^{12}C(d,p)^{13}C$ proved to be an important reference since its cross section has been widely measured, hence we were able to recognize its strong contribution in our experiment thanks to the measurements made by Kashy *et al.* at 165° in the incident energy range of $E_d$= [802-1956] keV [10], Jarjis *et al.* at 135° [11] within $E_d$=[2700- 2960] keV and at $E_d$=[3211-4333] keV at 150° measured by Bonner *et al.* [12]. Regarding the exit channels for $^{13}C+d$, only $^{13}C(d,\alpha)^{11}B$ [13] and $^{13}C(d,p)^{14}C$ measured by Marion and Weber were available and thereby located, however, the cross section of the latter was measured at 135° [4] and therefore it was not possible to use it to fit our data. It is worth mentioning that the sum of the individual contributions should reproduce our experimental data, however, since some of the data available for the simulation are from experiments at angles different from 150° (this experiment), the strength is not reproduced. Therefore, those reactions that were present at backscattered angles different from 150° were used for identification only.

**Table 1.** Number of particles by unit solid angle (particles*sr) obtained by SIMNRA and the total number of particles calculated by the solid angle

| Sample | Particles*sr | No. particles |
|---|---|---|
| **C2** | 9.00E+13 | 1.70E+17 |
| **C3** | 7.23E+13 | 1.27E+17 |

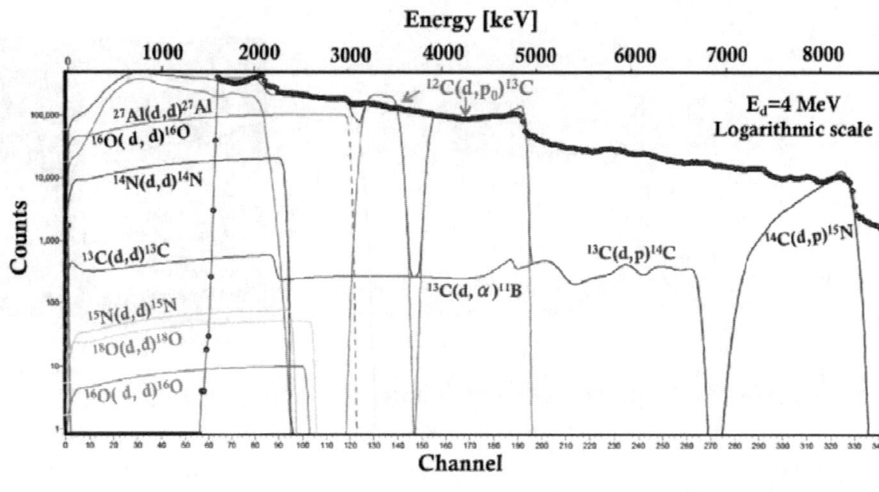


**Figure 2.** Backscattering protons spectrum obtained (shown is dark red points connected by a red line to guide the eye) set in logarithmic scale, compared with the simulated contributions from individual reactions performed with SIMNRA (shown in different colors).

The relative concentrations of $^{14}C/^{12}C$ and $^{13}C/^{12}C$ measured in the AMS system at the LEMA facilities are summarized in Table 2, where a clear difference of one order of magnitude can be seen between the irradiated samples and the non-irradiated ones, which demonstrates the production of $^{14}C$. Also there is an important difference in the relative concentrations for C2 and C3. This can be attributed in part to the different incident deuteron flux received by the samples (Table 1). Other effects like the ratio of the irradiated area in the graphite surface (1.33 $mm^2$) and the area of the deuteron beam ($\sim$ 2.69 $mm^2$) might also influence the measured concentration and further studies are required for this.

IOP Conf. Series: Journal of Physics: Conf. Series **1078** (2018) 012015 doi:10.1088/1742-6596/1078/1/012015

**Table 2.** Relative concentrations of $^{14}C/^{12}C$ and $^{13}C/^{12}C$

| Sample | | $^{14}C/^{12}C$ | $^{13}C/^{12}C$ |
|---|---|---|---|
| C2 | irradiated | 7.76E-12 ± 9.27E-19 | 1.43E-02 ± 6.59E-25 |
| C3 | irradiated | 1.53E-12 ± 3.10E-19 | 1.34E-02 ± 2.03E-25 |
| C4 | non-irradiated | 1.46E-13 ± 3.35E-19 | 1.31E-02 ± 7.31E-26 |
| C6 | non-irradiated | 1.72E-13 ± 3.73E-19 | 1.35E-02 ± 9.27E-26 |

Since the thickness of the targets were of 1.6 mm, an order of magnitude larger than the projected range of the deuterons going through graphite (165 $\mu$m at $E_{lab}$=4 MeV), the targets were treated as thick and therefore it was possible to compute the yields as follows:

$$y_i(E_0) = \frac{N_A f}{M_B} \int_{E_0}^{0} \sigma_i(E) \left[\frac{dE}{d(\rho x)}\right]^{-1} dE \qquad (1)$$

where $y_i$ corresponds to the yield of the reaction channel of interest, $N_A$ is the Avogadro's number, $f$ the isotopic fraction of the target nuclei of interest, $M_B$ the atomic weight of the target, $dE/d(\rho x)$ the stopping power of beam in the target per unit of mass thickness and $\sigma_i$ the cross section for the production for $i$.

For the nuclear reactions of interest, namely $^{12}C(d,p)^{13}C$ and $^{13}C(d,p)^{14}C$ we have correspondingly:

$$y_{13}(E_0) = \frac{N_A f_{12}}{M_B} \int_{E_0}^{0} \sigma_{13}(E) \left[\frac{dE}{d(\rho x)}\right]^{-1} dE \qquad (2)$$

and

$$y_{14}(E_0) = \frac{N_A f_{13}}{M_B} \int_{E_0}^{0} \sigma_{14}(E) \left[\frac{dE}{d(\rho x)}\right]^{-1} dE \qquad (3)$$

where $f_{12}$=0.989 and $f_{13}$=0.01 are the isotopic fractions for $^{12}C$ and $^{13}C$ respectively, $\sigma_{13}$ and $\sigma_{14}$ the cross sections for the production of $^{13}C$ and $^{14}C$.

Similarly, considering the reaction channels available at $E$ for $^{12}C$+$d$ and $^{13}C$+$d$ we will have the following equation:

$$Y_{12,13}(E_0) = \frac{N_A f_{12,13}}{M_B} \int_{E_0}^{0} \Sigma'_{12,13}(E) \left[\frac{dE}{d(\rho x)}\right]^{-1} dE \qquad (4)$$

where $\Sigma'_{12}$ and $\Sigma'_{13}$ are the sums of the reaction-channel cross sections of $^{12}C$+$d$ and $^{13}C$ +$d$ respectively, except for $(d,p)$ at $E$.

Likewise, we consider the yield for $^{13}C(d,p)^{14}C$ as $Y_{14} = N_{net14}/N_d$, where $N_{net14}$ is the difference between the $^{14}C$ present nuclei in the irradiated and non-irradiated samples; $N_d$, the number of incident deuterons. Thus, if there are $N_d$ incident projectiles, the total number of produced nuclei in the $i-th$ process will be $N_i = N_d Y_i$.

On the other hand, since several reactions take place simultaneously, the total number of particles of $^{12}C$, $^{13}C$ and $^{14}C$ will be correspondingly:

$$N_{12} = N_{12}^0 - N_d\, y_{13} - N_d \frac{N_A f_{12}}{M_B} \int_{E_0}^{0} \Sigma'_{12} \left[\frac{dE}{d(\rho x)}\right]^{-1} dE \qquad (5)$$

$$N_{13} = N_{13}^0 + N_d\, y_{13} - N_d\, y_{14} - N_d \frac{N_A f_{13}}{M_B} \int_{E_0}^{0} \Sigma'_{13} \left[\frac{dE}{d(\rho x)}\right]^{-1} dE \tag{6}$$

$$N_{14} = N_{14}^0 + N_d\, y_{14} - N_d \frac{N_A f_{14}}{M_B} \int_{E_0}^{0} \Sigma'_{14} \left[\frac{dE}{d(\rho x)}\right]^{-1} dE \tag{7}$$

From equations 5, 6 and 7 it is possible to obtain the expressions for $y_{13}$ and $y_{14}$ in terms of irradiated and non-irradiated relative concentrations ($N_{14}/N_{12}$, $N_{13}/N_{12}$ for the former, and $N_{14}^0/N_{12}^0$ and $N_{13}^0/N_{12}^0$ for the latter), the number of incident deuterons ($N_d$) and the yields for reaction channels other than $(d,p)$. These expressions provide a way to extract $y_{13}$ and $y_{14}$ from the experiment, with the exception of $Y_{12}$ and $Y_{13}$, which can be calculated from equation 4 using experimental cross sections reported in the literature [15]. Moreover the experimental results for $y_{13}$ and $y_{14}$ can be compared to those calculated using equations 2 and 3 and cross sections from the literature. Table 4 shows the experimental and calculated values of the yields.

$$y_{13} = \left[1 + \frac{N_{13}^0}{N_{12}^0} + \frac{N_{14}^0}{N_{12}^0}\right]^{-1} \cdot \left[\frac{N_{12}^0}{N_d}\left(\frac{N_{13}}{N_{12}} - \frac{N_{13}^0}{N_{12}^0} + \frac{N_{14}}{N_{12}} - \frac{N_{14}^0}{N_{12}^0}\right) + \right.$$
$$\left. + Y_{13} - \frac{N_{13}^0}{N_{12}^0} Y_{12} - \frac{N_{14}^0}{N_{12}^0} Y_{12}\right] \tag{8}$$

$$y_{14} = \frac{N_{12}^0}{N_d}\left(\frac{N_{14}}{N_{12}} - \frac{N_{14}^0}{N_{12}^0}\right) - \frac{N_{14}^0}{N_{12}^0}\left(\left[1 + \frac{N_{13}^0}{N_{12}^0} + \frac{N_{14}^0}{N_{12}^0}\right]^{-1} \cdot \right.$$
$$\left. \cdot \left[\frac{N_{12}^0}{N_d}\left(\frac{N_{13}}{N_{12}} - \frac{N_{13}^0}{N_{12}^0} + \frac{N_{14}}{N_{12}} - \frac{N_{14}^0}{N_{12}^0}\right) + Y_{13} - \frac{N_{13}^0}{N_{12}^0} Y_{12} - \frac{N_{14}^0}{N_{12}^0} Y_{12}\right] + Y_{12}\right) \tag{9}$$

**Table 3.** Yield calculation by two different ways. In the first column shows the yields derived using Tendl-2014 library data are enlisted. In the last column the values of the yields obtained using experimental data obtained in the present work

| | Yields using library data | Yields experimental data (this work) | |
|---|---|---|---|
| | | **C2** | **C3** |
| $Y_{12}$ | 2.05E-04 | - | - |
| $Y_{13}$ | 3.64E-05 | - | - |
| $y_{13}$ | 4.61E-04 | 4.67E-03 | 4.23E-03 |
| $y_{14}$ | 1.26E-02 | 3.57E-11 | 8.33E-12 |

Table 3 describes the yield values calculated using theoretical data (TENDL 2014) [14] retrieved from ENDF data base [15] and the values of the yields obtained using experimental data obtained in the present work, such as those enlisted in Table 2 and the number of incident deuterons in carbon, $N_d$. The significant disagreement between these values can be can be better explained observing Figure 3, which shows the comparison of $\Delta E \cdot E$ energy spectra for the concentration of $^{14}C$ between a non-irradiated sample (left) and an irradiated one (right). The dark-shaded area shown in both spectra is the region of interest (ROI), where the $^{14}C$ produced was expected to appear. However, as seen in the spectrum for the irradiated sample, a considerable amount of contamination can be clearly observed; on one hand, a significant

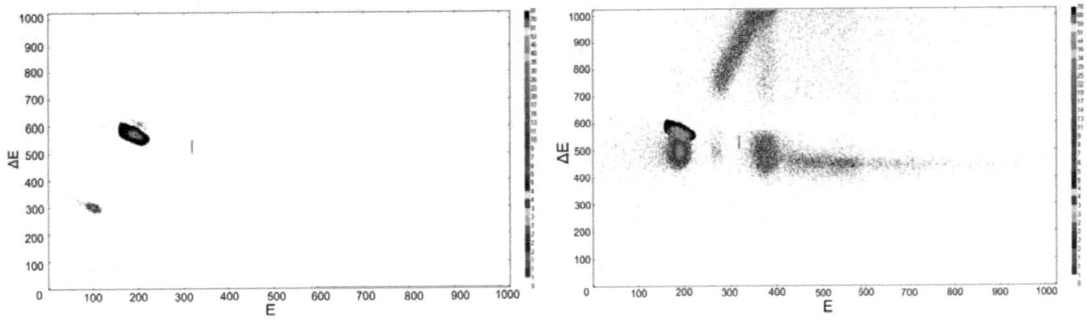


**Figure 3.** (left) $\Delta E \cdot E$ energy spectrum for the concentration of $^{14}C$ of a non-irradiated sample (C4);(right)$\Delta E \cdot E$ energy spectrum for the concentration of $^{14}C$ of an irradiated sample (C2).

disagreement between the TENDL data and the experimental data [4] [5] is consistent with this result (see Figure 4); on the other hand, a possible contamination with the $(^{12}C + {}^{2}H)^{++}$ molecular state (and maybe other contaminants) in our AMS measurements cannot be ruled out.

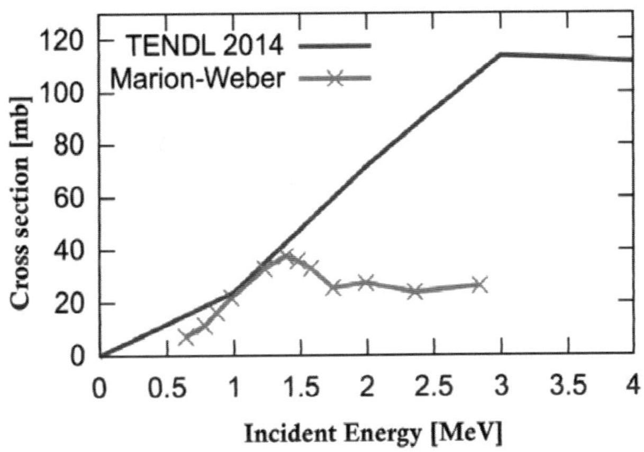


**Figure 4.** The red line shows the experimental data for the cross section of $^{13}C(d,p)^{14}C$ measured by Marion and Weber [4]. The blue line shows the cross section data of $^{13}C(d,p)^{14}C$ retrieved from TENDL 2014 library [14].

## 4. Conclusions

The production of $^{14}C$ was successfully achieved by performing the $^{13}C(d,p)^{14}C$ nuclear reaction using natural graphite as a target and 4 Mev deuterium beam. The AMS analysis performed at LEMA of the relative concentration of $^{14}C/^{12}C$ shows a significant amount of $^{14}C$ in the irradiated samples compared to those that did not undergo irradiation. The sample geometry used for the deuterium irradiation was that required at LEMA to enable direct AMS analysis after irradiation, however, this complicated the process because the graphite diameter was only 1.3 mm. On the other hand, the sample preparation implies working with thick targets, which also imposes difficulties in the extraction of cross sections of interest. A deconvolution of the yield is possible in order to extract the cross section as a function of energy. However, a series of measurements of the yield at different energies is needed. This work has helped to set the basis for future experiments using the AMS technique in the study of cross sections induced by

IOP Conf. Series: Journal of Physics: Conf. Series **1078** (2018) 012015 doi:10.1088/1742-6596/1078/1/012015

charged particles. Additional experiments and improvements, mainly in target preparation and beam focusing, are in order to complete a protocol to perform this type of measurements.

## Acknowledgments

The authors of this work are indebted to the 6.0 MeV Tandem Van de Graaff accelerator and LEMA technicians: Pedro Villaseñor and Sergio González. Thanks to Roxana Gutiérrez and Miguel Ángel Sánchez for their help at the Tandem accelerator at ININ. This work is possible only thanks to the financial support from CONACYT 271802 and 280760, DGAPAUNAM IG-101016.

## References

[1] Paul M. *et al.* (1980) *Phys. Lett.* B **94** 303-6
[2] Araujo-Escalona V. *et al.* (2017) *Phys. Proc.* **90** 421-28
[3] C. Solís *et al.* (2014) *Nuc. Inst. M. Phys.* **B** 233-37
[4] Marion J. and Weber G. (1956) *Phys. Rev.* **103** 167-71
[5] Colaux J., Thomé T. and Terwagne G. (2007)*Nucl. Instr. and Meth. in Phys. R.* B **254** 25-29
[6] Iocco F. *et al.* (2007) *Phys. Rev.* D **75** 087304
[7] Murillo-Morales S. *et al.* (2017) IOP *J. of Phys: Conf. S.* **876** 012017
[8] Mayer M. (1999) *AIP Conf. Proc.* **475** 541-44
[9] IBANDL database (2017) IAEA http://www-nds.iaea.org/ibandl/
[10] Kashy E. *et al.*(1960) *Phys. Rev.* **117** 1289
[11] Jarjis R. A. (1979) Int. Rep., U. of Manchester
[12] Bonner T.W. et al. (1956) Phys. Rev. **10** 209
[13] Marsh B. B. (1963) J. Phys. Rev. **130** 2373
[14] Koning A and Rochman D (2014) TENDL-2014 *Nuclear Research and Consultancy Group (NRG) Petten, The Netherlands.*
[15] V.Zerkin (2013) IAEA, https://www-nds.iaea.org/exfor/endf.htm

IOP Conf. Series: Journal of Physics: Conf. Series **1078** (2018) 012016 doi:10.1088/1742-6596/1078/1/012016

# Clustering phenomena: perspectives from iThemba LABS

**R. Neveling**
iThemba Laboratory for Accelerator Based Sciences, Somerset West 7129, South Africa
E-mail: neveling@tlabs.ac.za

**Abstract.** Aspects of alpha-clustering in nuclei were studied at iThemba LABS over the past twenty years by means of quasi-free scattering, inelastic scattering and transfer reactions and through particle decay spectroscopy. An overview of relevant work is provided, which highlights the unique opportunities at iThemba LABS, specifically with the K600 zero degree facility and recently commissioned ancillary detection systems.

## 1. Introduction

The study of clustering phenomena in nuclei, in particular alpha-clustering, is almost as old as the field of nuclear physics itself, and essentially started when Rutherford first distinguished between $\alpha$ and $\beta$ rays in the radiation emitted from uranium and its compounds [1]. These rays were later identified as particles, specifically the core of the $^4$He atom (or $\alpha$ particles) and electrons. The $\alpha$-rays Rutherford observed were $\alpha$-particles undergoing quantum-mechanical tunneling from inside the decaying nucleus. The question whether these $\alpha$ particles could be considered as pre-formed inside the nucleus ushered in a research topic on clustering that is still of interest today.

Various aspects of clustering phenomena have been under investigation at the iThemba Laboratory for Accelerator Based Sciences (iThemba LABS), South Africa. The majority of these exploit the capabilities of the K600 magnetic spectrometer at iThemba LABS, a high resolution kinematically corrected quadrupole dipole dipole magnetic spectrometer for light ions. In recent years the capability to measure inelastically scattered particles and reactions at extreme forward angles that includes zero degrees was successfully developed [2], making it one of only two facilities worldwide (the other being at RCNP, Japan [3]) where high energy resolution is combined with zero degree measurements at medium beam energies. The advantage of such measurements is the selectivity it provides to excitations with low angular momentum transfer. For this reason it represents a valuable experimental tool for research into E0 and E1 resonance strengths [4, 5, 6] but crucially also in finding and identifying $\Delta$L=0 cluster states. The recent addition of coincident particle and gamma detection capabilities [7] further enhances the selectivity of such a facility, which opens up a host of new opportunities to be explored.

## 2. Clustering in the ground state

The focus of studies into alpha clustering phenomena is commonly in relation to the appearance of cluster structures at and beyond excitation energies equivalent to alpha particle decay thresholds. This state of affairs follows from the definition for the emergence of clustering phenomena as developed by Ikeda *et al.* [8]. According to this picture the $^8$Be nucleus is the only example whereby it would be reasonable to expect alpha clusters appearing in the ground state, as all states in $^8$Be lies above the alpha decay threshold. However, anti-symmetrized molecular dynamics (AMD) calculations [9] indicate a non-negligible cluster structure in the mass density distributions for the ground state in the case of the $^{12}$C nucleus. It would therefore seem that the concept of ground state clustering is not entirely unreasonable.

The most direct experimental method to test the notion of ground state $\alpha$-clustering is by means of a knockout reaction in which the knocked out cluster is observed in coincidence with the projectile. The assumption that clusters of nucleons exist in nuclei would be strongly supported if the momentum distribution of the clusters as inferred from the coincidence spectra of emitted particles is in agreement with that expected from a preformed cluster bound in a target nucleus. This follows from good shape agreement between distorted wave impulse approximation (DWIA) [10] calculations and experimental energy sharing differential cross section data, as well as agreement between extracted spectroscopic values and theoretical predictions thereof.

However, DWIA cross sections are sensitive to inaccuracies in the often poorly known optical potentials used to generate wave functions of composite particles with nuclei, thus affecting the absolute value [11] and shape of the calculated cross sections. Fortunately for experiments using polarized beams there is an observable, the analyzing power $A_y$, that can act as a more stringent test of the reaction dynamics [12], which directly affects the interpretation of the results in terms of the existence of preformed $\alpha$ clusters in the ground state. If the reaction dynamics can be shown to be a quasi-free knockout mechanism, one expects that the analyzing power at the quasi-free peak will correspond to that of free $p+\alpha$ elastic scattering. The analyzing power is in essence a ratio of cross-sections, and is therefore in general less sensitive to the ingredients of the DWIA calculations, especially for light nuclei where there is minimal distortion of the incoming and outgoing waves.

This was indeed shown to be the case when the analyzing power distribution in the $^{12}$C$(p,p\alpha)^8$Be(g.s.) reaction was studied at iThemba LABS at an incident energy of 100 MeV [13]. The experimental analyzing power of the coincidence measurement was found to be in good agreement with the DWIA prediction as well as the trend of the analyzing power of free $p+^4$He elastic scattering, as shown in Fig.1. The results are consistent with a projectile interacting with $\alpha$ clusters as if they are free entities inside the nucleus, with the rest of the target nucleus acting merely as a spectator, and can be considered as evidence for the existence of preformed $\alpha$-clusters in the ground-state of $^{12}$C. Investigations of the $(p,p\alpha)$ reaction on the heavier $^{40}$Ca target with the K600 in coincidence with a single Si surface barrier detector telescope reveal a bigger influence of the distorting potentials on the analyzing power distributions, such that the results no longer agrees with free p+$^4$He elastic scattering [14]. However, proper agreement between the experimental analyzing power distribution and DWIA calculations can still be obtained provided the $\alpha+^{36}$Ar system is properly treated, which is indicative of the existence of a ground-state $\alpha$-cluster structure in $^{40}$Ca [15].

Recent interest in the use of quasi-free knockout reactions to investigate clustering in even heavier nuclei follows from theoretical calculations by Typel [16] describing the formation of $\alpha$-clusters on the surface of Sn itopes. While plans are afoot to use the $(p,p\alpha)$ reaction on $^{112-124}$Sn at 300 MeV at RCNP to investigate these claims, Cowley [17] used existing experimental information from $(d,^6Li)$ cluster pickup reactions to verify the predicted trend. The validity of the substitution of the $(d,^6Li)$ reaction for the $(p,p\alpha)$ knockout reaction follows from work by Carey *et al.* [18] who found a remarkable similarity of the relative spectroscopic factors extracted from $(p,p\alpha)$ and $(d,^6$Li) reactions for target nuclei covering a large mass range across the periodic table.

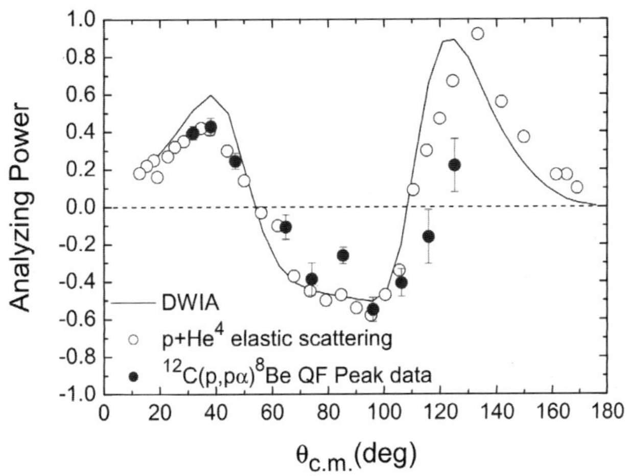


**Figure 1.** Experimental analyzing power distribution for the $^{12}C(p, p\alpha)^8Be$ reaction (filled circles) compared with a DWIA calculation (solid line) and free $p+\alpha$ elastic scattering data (empty circles) displayed as a function of the two-body center-of-mass p-$\alpha$ scattering angle, from Ref. [13].

## 3. $2^+$ excitation of the Hoyle state

Whereas quasi-free knockout reactions do not comment on the structural arrangement of clusters inside the nucleus, such information can be deduced from collective excitations of cluster states. The famous Hoyle state, the $0_2^+$ state at 7.654 MeV in $^{12}C$, is widely considered to have a $3\alpha$-cluster structure [19, 20]. The first potential $2^+$ state above the Hoyle state is reported to be at 11.160 MeV [21]. This state was observed in one experimental measurement only, namely in the investigation of the $^{11}B(^3He, d)$ reaction at 44 MeV with the Michigan 180° double-focussing magnetic spectrometer [22]. Its position is problematic, as at more then 3.5 MeV above the Hoyle state this $2^+$ candidate does not agree with predictions for a linear chain arrangement for the Hoyle state, which would imply a collective $2^+$ state at ~0.8 MeV above the Hoyle state [23]. A Hoyle state consisting of a dilute self-bound gas of bosonic $\alpha$-particles would, from BEC calculations, locate a 2+ state at ~ 1.3 MeV above the Hoyle state [24]. Close, but still not quite there, FMD calculations locate the first 2+ state at ~ 2.3 MeV above the Hoyle state [25].

In fact, no clear peak was observed by Reynolds *et al.* [22], and the $2^+$ assignment for the spin and parity of the state is only tentative. To date this observation remains unconfirmed. It is worthwhile to mention that the original $^{11}B(^3He, d)$ measurement was performed with photographic plates at the focal plane. The linear response of such detectors are typically superior to the non-uniform effects seen with the Multi Wire Drift Chambers (MWDC) used in more modern focal plane detectors. However, the event-by-event data reconstruction capabilities of modern data acquisition systems allow for a more thorough investigation of beam background and target contamination sources. In order to allow for a definitive assignment of spin and parity to this state, the measurement was revisited. The repeat of the $^{11}B(^3He, d)^{12}C$ measurement was performed at iThemba LABS with the K600 magnetic spectrometer. No evidence was found for the previously reported $2^+$ state at 11.16 MeV [26], and it was not possible to identify contaminating candidates responsible for the strength observed around 11.16 MeV in Ref. [22],

IOP Conf. Series: Journal of Physics: Conf. Series **1078** (2018) 012016 doi:10.1088/1742-6596/1078/1/012016

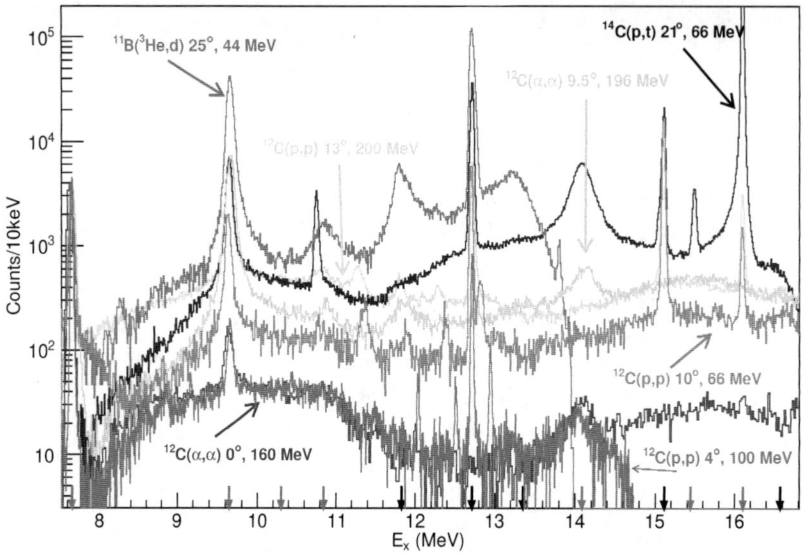


**Figure 2.** Excitation energy spectra of the $^{12}$C nucleus as obtained from various experiments performed at iThemba LABS with the K600 [26, 27, 31, 32, 33]. The different datasets were normalized on the strength of the Hoyle state. The arrows on the horizontal axis indicate the position of known excited states in $^{12}$C (but excluding the 11.16 MeV $2^+$ state [26] and the 10.13 MeV $2^+$ state [28]). Red arrows indicate natural parity states. Contaminant peaks from other carbon isotopes as well as oxygen are especially evident in the proton inelastic scattering data.

such as different particles with the same rigidity emitted as reaction particles from possible $^{10}$B, O, C, or N contaminants in the target of Ref [22].

Without the presence of the tentatively assigned $2^+$ state at 11.16 MeV the question is: where is the $2^+$ excited state of the Hoyle state? The first strong evidence for the existence of a new candidate $2^+$ excitation of the Hoyle state was provided by inelastic proton scattering measurements at iThemba LABS. The high sensitivity experiment was performed with a 66 MeV proton beam, observing the scattered protons with the K600 magnetic spectrometer at a range of angles chosen to emphasize the $2^+$ excitation while suppressing the other $^{12}$C excited states. Evidence was found for a possible $2^+$ state at 9.6(1) MeV with a width of 600(100) keV [27]. The state in question falls within a region dominated by other states, some of which are very broad, such as the broad $0^+$ strength around 10 MeV. This makes the $2^+$ excited state of the Hoyle state, itself a broad state, very difficult to observe. This can be seen in Fig.2 in the typical excitation energy spectra of $^{12}$C obtained with various probes. The definitive experiment establishing the position and width of the $2^+$ excited state of the Hoyle state at $E_x$=10.03(11) MeV and $\Gamma$=800(130) keV respectively was the measurement of the $^{12}C(\gamma, 3\alpha)$ reaction at HI$\gamma$S [28]. The major advantage of this measurement was the fact that $\gamma$-ray beams cannot populate $0^+$ states and only populates the $3^-$ state weakly. Therefore it does not suffer from the influence of the overwhelming background conditions as in the case of the inelastic proton scattering experiment at iThemba LABS.

IOP Conf. Series: Journal of Physics: Conf. Series **1078** (2018) 012016 doi:10.1088/1742-6596/1078/1/012016

## 4. The search for $0^+$ states

The exact nature of the structure of the Hoyle state, which is widely considered to have a $3\alpha$-cluster structure [19, 20], remains unresolved. In one interpretation it is considered to have a $3\alpha$ gas-like structure similar to a Bose-Einstein condensate consisting of three $\alpha$ particles all occupying the lowest $0S$ state [29], although it has also been shown that one must be careful when referring to only three particles being in a Bose-Einstein condensation [25]. Despite the fact that the details of the structure of the Hoyle state are still debated, it is widely considered to be a prototype for a class of states near the $\alpha$-particle thresholds in light self-conjugate 4n nuclei [25, 29]. In the spirit of this conjecture the neighbouring $^{16}$O and $^{20}$Ne nuclei were experimentally investigated at iThemba LABS.

A potential Hoyle-like candidate in $^{16}$O was identified by Funaki *et al.* [30]. They solved a four-body equation of motion based on the Orthogonality Condition Model (OCM) that succeeded in reproducing the observed $0^+$ spectrum in $^{16}$O up to the $0_6^+$ state. It was shown that a $4\alpha$ condensation state could be assigned to the $0_6^+$ state located at 15.097 MeV, for which the calculation yields a large radius of 5 fm, indicating a dilute density structure. It is in principle possible to gain insight into the structure of the state by studying its the decay properties. The decay of the Hoyle state is known to take place mostly through $^8$Be+$\alpha$ decay, with the upper limit for direct 3-$\alpha$ decay known to be in the range 0.043%-0.047% [34, 35]. Although this makes it tempting to simplistically claim a $^8$Be+$\alpha$ structure for the Hoyle state, it was found that the observed decay is consistent with theoretical calculations that assumes an explicit $3\alpha$ structure [36]. In order to investigate the particle decay of the Hoyle-like candidate in $^{16}$O the $^{16}\mathrm{O}(\alpha,\alpha')$ reaction was studied at 0° with the K600 magnetic spectrometer, with coincident proton and $\alpha$-decay from the natural parity states observed in a large-acceptance silicon-strip detector array called CAKE at backward angles [37], as shown in Fig.3. Surprisingly, instead of enabling the characterization of the decay path of the $0_6^+$ state at $E_x = 15.097$ MeV, the angular correlations of charged-particle decays from $^{16}$ indicate the presence of an unresolved resonance at $E_x \sim 15$ MeV with non-zero spin [38]. This might explain the disparity between the theoretical and experimentally observed widths of 34 keV [39] and 166(30) keV, respectively. Clearly more detailed investigations with higher energy resolution should be performed into this excitation energy region in future.

In the case of the $^{20}$Ne nucleus no known candidate $0^+$ state exists near the $5\alpha$ breakup threshold at $E_x = 19.17$ MeV. There are 49 known states in this nucleus at excitation energies above 18.43 MeV [40], none of which have spin values below J = 4. The excitation energy spectrum of the $^{20}$Ne nucleus, from the ground state to $E_x$= 25 MeV, was subsequently investigated with the $^{22}\mathrm{Ne}(p,t)^{20}\mathrm{Ne}$ reaction at a beam energy of 60 MeV with the K600 magnetic spectrometer at numerous angles, including zero degrees [41]. Narrow, low-spin $J^\pi = 0^+$,1, and $2^+$ states have been observed for the first time at high-excitation energies in $^{20}$Ne, namely at $E_x$= 17.67, 18.84, 20.59, 21.16, 21.80, and 22.5 MeV. Shell-model and coupled reaction channel calculations were performed to determine the nature of the newly observed states. The state at 22.5 MeV, which could not be interpreted by the shell-model calculations, can be considered a tentative candidate for the $5\alpha$ cluster state.

More generally, clustering phenomena can be associated with large deformation in nuclei, and associated superdeformed rotational bands have been observed in nuclei such as $^{36}$Ar and $^{40}$Ca [42]. In lighter nuclei such as $^{28}$Si the superdeformed bands are less clear, since the standard techniques to observe these bands rely on heavy ion fusion evaporation reactions in order to populate high spin states, followed by the observation of $\gamma$-rays from decays down the band. Such methods are not efficient in observing the low-spin states. Of particular importance is locating the $0^+$ bandhead of previously tentatively identified superdeformed bands. For this reason the K600 magnetic spectrometer was used for inelastic $\alpha$-particle scattering on $^{28}$Si at very forward angles, to allow for the clear identification of $0^+$ states. A new $0^+$ state at 9.71 MeV

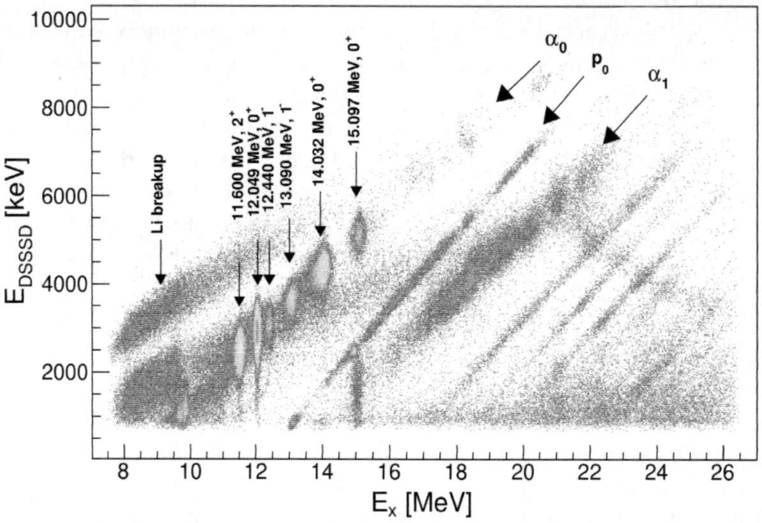


**Figure 3.** Two-dimensional coincidence matrix for inelastically scattered $\alpha$-particles from a $^{nat}Li_2CO_3$ target summed over all angles in the silicon-strip detector array. Three $^{16}O$ decay channels ($\alpha_0$, $\alpha_1$, $p_0$) are indicated, as well as prominent low spin states in $^{16}O$, including the supposedly $0_6^+$ state at $E_x = 15.097$ MeV. From Ref.[44].

was subsequently identified [43], which is a strong candidate for the bandhead of a previously identified superdeformed band.

## 5. Future work

The combination of good energy resolution through dispersion matching techniques and the zero degree capability of the K600 makes it possible to search for cluster states through low spin transfer reactions such as $(p,t)$ and $(\alpha,\alpha')$. While the experimental studies so far focused on self-conjugate 4n nuclei, it is also envisaged to investigate matters relating to the $2\alpha + t$ cluster structure in $^{11}B$, where theoretical calculations [45] indicate many positive parity states around the $2\alpha + t$ threshold with features of developed clusters. Finally, it is foreseen that the polarized ion source at iThemba LABS will be put to good use in inclusive analyzing measurements to further investigations into the ground-state clustering of heavy nuclei.

## Acknowledgements

The financial assistance of the South African National Research Foundation (NRF) towards research at iThemba LABS is hereby acknowledged. R.N. specifically acknowledges financial support from the NRF through Grant No. 85509. The results presented here represent the work of various collaborations within the K600 working group.

## References

[1] E. Rutherford, Philosophical Magazine, January 1899, ser. 5, xlvii, 109.
[2] R. Neveling *et al.*, Nucl. Instr. and Meth. A 654 (2011) 29.
[3] A. Tamii *et al.*, Nucl. Instr. and Meth. A 605 (2009) 326.
[4] A. Tamii *et al.*, Phys. Rev. Lett. 107 (2011) 062502.
[5] M. Itoh *et al.*, Phys. Rev. C 88 (2013) 064313.

[6] A.M. Krumbholz *et al.*, Phys. Lett. B 744 (2015) 7.
[7] R. Neveling *et al.*, Proceedings of the $61^{st}$ SAIP conference, edited by S Peterson and S Yacoob, The South African Institute of Physics, Cape Town (2017) 134.
[8] K. Ikeda, N. Tagikawa and H. Horiuchi, Prog. Theo. Phys. Suppl. E68 (1968) 464.
[9] Y. Kanada-En'yo, Prog. Theor. Phys. 117 (2007) 655.
[10] N.S. Chant and P.G. Roos, Phys. Rev. C 27 (1983) 1060.
[11] P.E. Hodgson and E. Běták, Phys. Rep. 374 (2003) 1.
[12] R. Bonetti, F. Crespi, and K.-I. Kubo, Nucl. Phys. A 49 (1989) 381.
[13] A.A. Cowley *et al.*, EPL, 85 (2009) 22001; J Mabiala *et al.*, Phys. Rev. C 79 (2009) 054612.
[14] R. Neveling *et al.*, Phys. Rev. C 77 (2008) 037601.
[15] A.A. Cowley, Journal of Physics: Conference Series 436 (2013) 012011.
[16] S. Typel, Phys. Rev. C 89 (2014) 064321.
[17] A.A. Cowley, Phys. Rev. C 93 (2016) 054329.
[18] T.A. Carey *et al.*, Phys. Rev. C 29 (1984) 1273.
[19] M. Freer, H. Horiuchi, Y. Kanada-En'yo, D. Lee, Ulf-G. Meissner, arXiv:1705.06192.
[20] M. Itoh *et al.*, Journal of Physics: Conference Series 863 (2017) 012019.
[21] F. Ajzenberg-Selove and J.H. Kelley, Nuclear Physics A506 (1990) 1,
[22] G.M. Reynolds, D.E. Rundquist and M. Poichar, Phys. Rev. C3 (1971) 442.
[23] A. C. Merchant *et al.*, Nucl. Phys. A 549 (1992) 431.
[24] Y. Funaki *et al.*, Phys. Rev. C 67 (2003) 051306.
[25] M. Chernykh *et al.*, Phys. Rev. Lett. 98 (2007) 032501.
[26] F.D. Smit *et al.*, Phys. Rev. C 86 (2012) 037301.
[27] M. Freer *et al.*, Phys. Rev. C 80 (2009) 041303(R).
[28] W.R. Zimmerman *et al.*, Phys. Rev. Lett. 110 (2013) 152052.
[29] P. Schuck *et al.*, Journal of Physics: Conference Series 413 (2013) 012009.
[30] Y. Funaki *et al.*, Phys. Rev. Lett. 101 (2008) 082502.
[31] F. D. Smit *et al.*, Project PR166, Research Proposal to the PAC of iThemba LABS, September 2009.
[32] F. Nemulodi *et al.*, Project PR195, Research Proposal to the PAC of iThemba LABS, September 2011.
[33] R. Neveling *et al.*, Project PR194, Research Proposal to the PAC of iThemba LABS, September 2011.
[34] R. Smith *et al.*, Phys. Rev. Lett. 119 (2017) 132502.
[35] D. Dell'Aquila *et al.*, Phys. Rev. Lett. 119 (2017) 132501.
[36] S. Ishikawa, Phys. Rev. C 90 (2014) 061604(R).
[37] P. Adsley *et al.*, JINST 12 (2017) T02004
[38] K.C.W. Li *et al.*, Phys. Rev. C 95 (2017) 031302(R).
[39] Y. Funaki *et al.*, Phys. Rev. C 80 (2011) 064326.
[40] Experimental Nuclear Reaction Data Library (EXFOR), IAEA Nuclear Data Section. See http://www-nds.iaea.org/exfor/ or, for the NNDC at Brookhaven National Laboratory, the mirror site is http://www.nndc.bnl.gov/exfor/.
[41] C. Swartz *et al.*, Phys. Rev. C 91 (2015) 034317.
[42] E. Ideguchi *et al.*, Phys. Rev. Lett. 87 (2001) 222501.
[43] P. Adsley *et al.*, Phys. Rev. C 95 (2017) 024319.
[44] R. Neveling *et al.*, Proceedings of the 14th International Conference on Nuclear Reaction Mechanisms, Varenna, Italy, 15-19 June 2015, CERN-Proceedings-2015-001 (2015) 79.
[45] B. Zhou and M. Kimura, arXiv:1711.04439v1.

IOP Conf. Series: Journal of Physics: Conf. Series **1078** (2018) 012017 doi:10.1088/1742-6596/1078/1/012017

# Probing the Early Universe through nuclear physics

**R.G. Pizzone[(a)], R. Spartá[(a)], C.A. Bertulani[(c)], M. La Cognata[(a)], L. Lamia[(a,b)], C. Spitaleri[(a,b)], A. Tumino[(a,d)]**

[(a)]INFN Laboratori Nazionali del Sud, Catania, Italy
[(b)]Dipartimento di Fisica e Astronomia, Universitá di Catania, Catania, Italy
[(c)] Texas A&M University, Commerce, Texas, USA
[(d)]Universitá Kore di Enna, Enna, Italy

E-mail: rgpizzone@lns.infn.it

**Abstract.** Big Bang Nucleosynthesis (BBN) requires several nuclear physics inputs and nuclear reaction rates. An up-to-date compilation of direct cross sections of d(d,p)t, d(d,n)$^3$He and $^3$He(d,p)$^4$He reactions is given, being these ones among the most uncertain bare-nucleus cross sections. An intense experimental effort has been carried on in the last decade to apply the Trojan Horse Method (THM) to study reactions of relevance for the BBN and measure their astrophysical S(E)-factor. The reaction rates and the relative error for the four reactions of interest are then numerically calculated in the temperature ranges of relevance for BBN ($0.01<T_9<10$). These value were then used as input physics for primordial nucleosynthesis calculations in order to evaluate their impact on the calculated primordial abundances and isotopical composition for H, He and Li. New results on the $^7$Be(n,$\alpha$)$^4$He reaction rate were also taken into account.These were compared with the observational primordial abundance estimates in different astrophysical sites. Reactions to be studied in perspective will also be discussed

## 1. Introduction

One of the foundation stones of the Big Bang model, together with the Hubble expansion and the Cosmic Microwave Background (CMB) radiation [1] is the Big Bang Nucleosynthesis (BBN). BBN probes the Universe to the earliest times, the so called radiation dominated era, from a fraction of second to few minutes. It involves events that occurred at temperatures below 1 MeV, and naturally plays a key role in forging the connection between cosmology and nuclear physics [2]. Focusing only on the products of the BBN, according to the Standard Big Bang Nucleosynthesis model (SBBN), only the formation of light nuclei ($^2$H,$^{3,4}$He,$^7$Li) is predicted in observable quantities, starting from protons and neutrons. Today, with the only exception of $^3$He and lithium, the abundances of these isotopes in the appropriate astrophysical environments are rather consistent with SBBN predictions [3]. A comparison between the primordial abundances deduced from WMAP CMB precise measurements and the calculated ones constrains the baryon-to-photon ratio, $\eta$, which is the only free parameter in the presently accepted model of the SBBN. A recent observation yields $\eta = 6.16 \pm 0.15 \times 10^{-10}$ [4], which is the value that we adopt in our calculations.

BBN nucleosynthesis requires several nuclear physics inputs and, among them, an important role is played by the nuclear reaction rates. Due to the relatively small amount of key nuclear species involved in the BBN nuclear reaction network, only 12 reactions play a major role [5].

IOP Conf. Series: Journal of Physics: Conf. Series **1078** (2018) 012017 doi:10.1088/1742-6596/1078/1/012017

Some of those reactions involve neutrons and radioactive ions and are currently not known with sufficient precision (see table 1).

| | | | |
|---|---|---|---|
| $n \leftrightarrow p$ (1) | $p(n,\gamma)d$ (2) | $d(p,\gamma)^3He$ (3) | $d(d,p)t^{(\dagger)}$ (4) |
| $d(d,n)^3He^{(\dagger)}$ (5) | $^3He(n,p)t$ (6) | $t(d,n)^4He$ (7) | $^3He(d,p)^4He^{(\dagger)}$ (8) |
| $^3He(\alpha,\gamma)^7Be$ (9) | $t(\alpha,\gamma)^7Li$ (10) | $^7Be(n,p)^7Li$ or $^7Be(n,\alpha)^4He^{(11)(\dagger)}$ | $^7Li(p,\alpha)^4He^{(\dagger)}$ (12) |

**Table 1.** Nuclear reactions of greatest relevance for Big Bang nucleosynthesis, labelled from 1 to 12. The reactions already measured with the Trojan Horse method are marked with a † symbol. Reaction (6) will be studied in a future experiment.

The reaction rates are calculated from the available low-energy cross sections for reactions which are also a fundamental input for a number of other still unsolved astrophysical problems, e.g. the so called "lithium depletion" either in the Sun or in other galactic stars [6, 7]. Cross sections should be measured in the astrophysically relevant Gamow window [8], of the order of few hundreds of keV. In the last decades these reactions have been widely studied and, in particular, great efforts have been devoted to their study by means of direct measurements at the relevant astrophysical energies, sometimes in underground laboratories [9, 10]. However, for many of the relevant reactions, no direct data exist at astrophysical energies (mostly because of difficulties connected with the presence of the Coulomb barrier in charged particle induced reactions) and the cross section within the Gamow window has to be extrapolated from higher energy measurements. Alternative and challenging ways to obtain the bare nucleus cross section, $\sigma_b$, for charged-particle reactions at sub-Coulomb energies have been provided by indirect methods such as the Coulomb dissociation method [11, 12] and the ANC (Asymptotic Normalization Coefficient) [13]. Among them, the Trojan-horse Method (THM) [14] is particularly suited to investigate binary reactions induced at astrophysical energies by neutrons or charged particles by using appropriate three-body reactions. It allows one to avoid both Coulomb barrier suppression and electron screening effects, thus avoiding the use of extrapolations. Moreover, it may be used with neutron induced reactions as well as radioactive isotopes, and even to determine cross sections of neutron induced reactions on unstable isotopes. The method has been used in the last three decades to explore nucleosynthesis associated with stellar phenomena, primordial nucleosynthesis as well as explosive nucleosynthesis. In the next sections we will show the calculations of the reaction rates based also on the THM measurements of $\sigma_b$. For recent reviews on the THM see [14]). Thus, the method can be regarded as a powerful indirect technique to get information on bare nucleus cross section for reactions of astrophysical interest, which leads to new reaction rate determinations. The method itself has been used to explore AGB stars nucleosynthesis [15, 16], novae [48] as well as LiBeB depletion in stars [17, 18, 19].

Many of the reactions relevant for the SBBN, i.e. $^7Li(p,\alpha)^4He$, $^2H(d,p)^3H$, $^2H(d,n)^3He$, $^3He(d,p)^4He$, $^7Be(n,\alpha)^4He$ were studied by means of the THM in the energy range of interest and their measurements were performed in an experimental campaign which took place in the last decade [20, 21, 32, 22, 23, 25, 27].

## 2. Reaction rates with TH data

The reaction rates for the the four reactions mentioned above (from a compilation of direct and THM data, as reported in [26]) have been calculated numerically. Then, we fitted the rates with the parametrization displayed in Equation 1. This is the common procedure adopted in previous works (see, e.g., [28, 29, 30]). For the 4 reactions of interest, we have fully included the experimental errors from measurements, allowing us to evaluate the respective errors in the

IOP Conf. Series: Journal of Physics: Conf. Series **1078** (2018) 012017 doi:10.1088/1742-6596/1078/1/012017

reaction rates. The numerical results are then fitted with the expression

$$N_A \langle \sigma v \rangle = \exp[a_1 + a_2 \ln T_9 + \frac{a_3}{T_9} + a_4 T_9^{-1/3} + a_5 T_9^{1/3} + a_6 T_9^{2/3} + a_7 T_9 + a_8 T_9^{4/3} + a_9 T_9^{5/3}] \quad (1)$$

which incorporates the temperature dependence (where $T_9$ is the temperature in GK) of the reaction rates during the BBN. The $a_i$ coefficients for the $^2$H(d,p)$^3$H and the $^2$H(d,n)$^3$He reactions are given for THM in Table 2, while the coefficients for the $^3$He(d,p)$^4$He and $^7$Li(p,$\alpha$)$^4$He reaction rate expression are given in Table 3. As regards the $^7$Be(n,$\alpha$)$^4$He reaction its reaction rate was calculated upon indirect measurements in [27].

| $a_i$ | $^2$H(d,p)$^3$H | $^2$H(d,n)$^3$He |
|---|---|---|
| $a_1$ | 14.996 | 16.1787 |
| $a_2$ | −2.4127 | −1.9372 |
| $a_3$ | $2.8261 \times 10^{-3}$ | $2.0671 \times 10^{-3}$ |
| $a_4$ | −5.3256 | −5.0226 |
| $a_5$ | 6.6125 | 5.7866 |
| $a_6$ | 2.4656 | $-2.039 \times 10^{-2}$ |
| $a_7$ | −3.8702 | −0.7935 |
| $a_8$ | 1.6700 | 0.2678 |
| $a_9$ | −0.25851 | $-3.1586 \times 10^{-2}$ |

**Table 2.** Table with reaction rate parameters (appearing in Eq. 1) for $^2$H(d,p)$^3$H and $^2$H(d,n)$^3$He evaluated from the present work.

| $a_i$ | $^3$He(d,p)$^4$He | $^7$Li(p,$\alpha$)$^4$He |
|---|---|---|
| $a_1$ | 20.4005 | 17.6686 |
| $a_2$ | 1.3850 | −1.1549 |
| $a_3$ | $-1.2982 \times 10^{-2}$ | $-4.4059 \times 10^{-4}$ |
| $a_4$ | −4.1193 | −8.5485 |
| $a_5$ | 12.2954 | 4.6683 |
| $a_6$ | −15.2114 | −0.7858 |
| $a_7$ | 5.4147 | −2.3208 |
| $a_8$ | −0.5048 | 2.0628 |
| $a_9$ | $-4.3372 \times 10^{-2}$ | −0.4747 |

**Table 3.** Table with reaction rate parameters (appearing in Eq. 1) for $^3$He(d,p)$^4$He and $^7$Li(p,$\alpha$)$^4$He evaluated from present work.

The direct data were considered from the compilation described in [26] for energies above 100 keV for $^3$He(d,p)$^4$He and $^7$Li(p,$\alpha$)$^4$He and for energies above 10 keV for $^2$H(d,p)$^3$H and $^2$H(d,n)$^3$He, in order to avoid the enhancement due to the electron screening in the direct data. TH data, which are not suffering the electron screening enhancement, essentially describe the bare nucleus contribution to the cross section.

For all the cases we noticed that deviations of up to 20% are obtained from previous compilations.

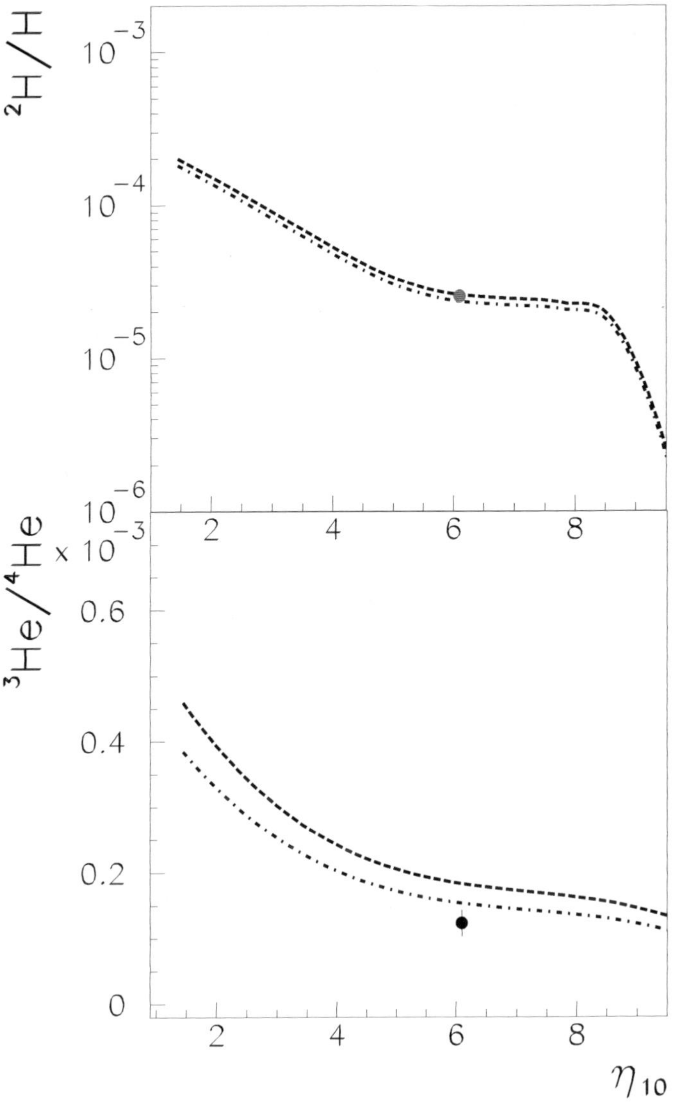


**Figure 1.** Primordial isotopic ratios of hydrogen (top) and He (bottom) isotopes as a function of the baryon-to-photon ratio $\eta_{10}$. The dashed lines represent the theoretical calculation uncertainty (upper and lower limits), as they arise from experimental errors. The red dot is the observed primordial value for D/H ratio ([42] ) while the black one the observed primordial value for $^3$He/$^4$He ([40] ).

## 3. Discussion and Perspectives

The reaction rates of four of the main reactions of the BBN network in the temperature range ($0.001 < T_9 < 10$), namely, $^2$H(d,p)$^3$H, d(d,n)$^3$He, $^3$He(d,p)$^4$He, $^7$Li(p,$\alpha$)$^4$He, have been calculated numerically including the recent THM measurements [31, 34, 35, 36, 38, 39, 24, 33, 37]. The uncertainties of experimental data for direct and THM data have been fully included for the above reactions. The extension of the same methodology to the other reactions forming the BBN

IOP Conf. Series: Journal of Physics: Conf. Series **1078** (2018) 012017 doi:10.1088/1742-6596/1078/1/012017

| Isotopic ratio | Pizzone et al. 2014 [26] | Observed |
|---|---|---|
| $^3$He/$^4$He | $(1.72\pm0.09)\times10^{-4}$ | $1.23\times10^{-4\ (a)}$ |
| D/H ($\times10^{-5}$) | $2.692^{+0.177}_{-0.070}$ | $2.82\pm0.26^{(b)}$ |
| $^7$Li/H | $4.683^{+0.335}_{-0.292}$ | $1.58\pm0.31^{(d)}$ |

**Table 4.** BBN predictions ($\eta_{10}$ =6.16) using different set of data (see text) compared with observations. (a) The observed $^3$He/$^4$He ratio is from Ref. [40]. (b) The mean observed deuterium abundance is the mean average from [42]. (d) The lithium abundance arises from observations of stars which provide a sample of the "lithium plateau" and is expressed in terms of Li/H in units $10^{-10}$.

reaction network will be examined in a forthcoming paper. The parameters of each reaction rates as given in Eq. 1 are reported in [26]. The obtained reaction rates are compared with some of the most commonly used compilations found in the literature and used to calculate the BBN abundance for $^{3,4}$He, D and $^7$Li. The obtained abundances are in agreement, within the experimental errors, with those obtained using the compilation of directly-measured reaction rates. Isotopical ratios D/H and $^3$He/$^4$He are plotted in figure 1 as a function of the baryon-to-photon ration $\eta$. Observative constraints are reported as solid marks in the figures for $\eta_{10} = 6.16 \pm 0.15$. Dashed lines represent the BBN model with the uncertainty related to the involved reaction rates. A comparison of our predictions with the observations for primordial abundance of D/H, $^3$He/$^4$He show an agreement as reported in table 4. The very-well-known *lithium problem*, i.e. the discrepancy between lithium abundance predicted by models and observed in primordial objects is still present.

The present results show the power of THM as a tool for exploring charged particle induced reactions at the energies typical of BBN. Further reactions to be explored by means of the THM include the $^3$He(n,p)$^3$H and the $^7$Be(n,p)$^7$Li reactions which are marked as (6) and (11) in table 1. This makes primordial nucleosynthesis the astrophysical scenario where THM has given the most contribution. The recent use of the method with neutrons induced reactions [45] as well as radioactive ion beams [46, 47] might provide a (possibly partial) nuclear solution to the long-standing lithium problem. However its definitive solution may lay in a better understanding of small-mass small metallicity stars.

## References

[1] G. Steigman 2007, Ann. Rev. Nucl. Part. Sci. 57, 463
[2] B. D. Fields and S. Sarkar 2006, J. Phys. G33, 220
[3] G. Israelian 2012, Nature 489, 37
[4] E. Komatsu et al. 2011, Ap. J. Sup. 192, 18
[5] E.W. Kolb and M.S. Turner 1990, "The Early Universe", Addison-Wesley
[6] R. Weymann and E. Moore 1963, Ap. J. 137, 552
[7] D. Ezer and A.G.W. Cameron 1963, Icarus 1, 422
[8] C. Iliadis 2007, "Nuclear Physics of Stars", Wiley
[9] R. Bonetti et al. 1999, Phys. Rev. Lett. 82, 5205
[10] C. Casella et al. 2002, Nucl. Phys. A706, 203
[11] G. Baur, C.A. Bertulani & H Rebel 1986, Nucl. Phys. A458, 188
[12] C.A. Bertulani & A. Gade 2010, Phys. Rep. 485, 195
[13] A. Mukhamezhanov et al. 2008, Phys. Rev. C 78, 0158042008
[14] C. Spitaleri et al. 2016, Eur. Phys. J. A**52**, 77
[15] S. Palmerini, M.L Sergi, M. La Cognata, L. Lamia, et al., 2013, Ap. J., 764, 128
[16] R.G. Pizzone, G.D'Agata, C. Spitaleri et al., 2017, Ap. J., 836, 57
[17] C. Spitaleri, L. Lamia, et al., 2014 Phys. Rev. C 90, 035801
[18] M. Aliotta, C. Spitaleri, et al., 2000. Eur. Phys. J. 9, 435
[19] L. Lamia et al., 2008, Nuovo Cimento C, 31, 423-431
[20] A. Tumino et al. 2008, Phys. Rev. C 78, 064001
[21] R.G. Pizzone et al. 2005, A. & A. 438, 779
[22] L. Lamia et al., 2013, Ap. J. 768, 65
[23] L. Lamia, et al. 2012, Phys. Rev. C 85, 025805
[24] R.G. Pizzone et al. 2011, Phys. Rev. C, 83, 045801
[25] L. Lamia et al. 2015, Ap.J., 811, 99
[26] R.G. Pizzone et al. 2014, Ap. J.,786, 112
[27] L. Lamia et al. 2017, Ap. J. 850, 175
[28] M.S. Smith, L.H. Kawano and R.A. Malaney 1993, Ap. J. 85, 219
[29] R.H. Cyburt 2004, Phys. Rev. D 70, 023505
[30] A. Coc, S. Goriely, Y. Xu, M. Saimpert, and E. Vangioni 2012, Astrophys. J. 744, 18
[31] L. Lamia et al. 2012, A. & A. 541, 158
[32] A. Rinollo et al. 2005, Nucl. Phys. A 758, 146c
[33] M. Lattuada et al. 2001, Ap. J. 562, 1076
[34] A. Tumino et al. 2011, Phys. Lett. B 705 (5), 546
[35] A. Tumino et al. 2014, Ap. J., 785, 96
[36] M. La Cognata et al. 2005, Phys. Rev. C 72, 065802
[37] R.G. Pizzone et al. 2003, A. & A. 398, 423
[38] R.G. Pizzone et al. 2013, Phys. Rev. C 87, 025805
[39] C. Li et al., 2015, Phys. Rev. C, 92, 025805
[40] H. Busemann et al., 2001, L & PS, 32, 1598
[41] Y.I.Izotov and T.X.Thuan 2010, Ap. J. Lett. 710, L67
[42] J.M. O'Meara, S. Burles, J.X. Prochaska, and G.E. Prochter 2006, Ap. J. 649, L61
[43] T. M. Bania, R. T. Rood, D. S. Balser 2002, Nature 415, 54
[44] L. Sbordone, et al. 2010, Astron. Astro. 522, 26
[45] M. Gulino et al. 2013, Phys. Rec. C87, 012801
[46] S. Cherubini, et al. 2015, Phys. Rev. C 92, 015805
[47] R.G. Pizzone et al. 2016, Eur. Phys. J. A52, 24
[48] M. La Cognata, R.G. Pizzone et al., 2017, Ap. J., 846, 65

IOP Conf. Series: Journal of Physics: Conf. Series **1078** (2018) 012018 doi:10.1088/1742-6596/1078/1/012018

# B-L extension of the SM - $Z'$ and heavy Higgs boson analysis at future $e^+e^-$ linear colliders ILC and CLIC

**F Ramírez-Sánchez[1], A Gutierrez-Rodríguez[2] and M A Hernández-Ruiz[3]**

[1] Facultad de Física, Universidad Autónoma de Zacatecas, Apartado Postal F-630, 99000 Fresnillo, Zac. México.
[2] Facultad de Física, Universidad Autónoma de Zacatecas, Apartado Postal C-580, 98060 Zacatecas, Zac. México.
[3] Facultad de Ciencias Químicas, Universidad Autónoma de Zacatecas, Apartado Postal C-580, 98060 Zacatecas, Zac. México.

E-mail: [1]frank.ramirez@fisica.uaz.edu.mx, [2]alexgu@fisica.uaz.edu.mx and [3]mahernan@uaz.edu.mx

**Abstract.** We propose a $U(1)_{\rm B-L}$ (baryon minus lepton) extension to the Standard Model, which adds a complex singlet to the scalar sector where an extra neutral gauge boson $Z'$ corresponding to the $U(1)_{\rm B-L}$ gauge symmetry and an extra SM like singlet scalar $H$ (heavy Higgs) are predicted. We consider the mixture of the $Z$ and $Z'$. The model also yields Majorana masses for three right handed neutrinos. We study the phenomenology of the light $h$ and heavy $H$ Higgs boson production and decay at future $e^+e^-$ linear colliders with center-of-mass energies of $\sqrt{s} = 500 - 3000$ GeV and integrated luminosities of $\mathcal{L} = 500 - 2000$ fb$^{-1}$. The study includes the Higgs-strahlung processes $e^+e^- \to (Z\,Z') \to Zh$, $e^+e^- \to (Z\,Z') \to ZH$, and we complement our previous works with the addition of the $e^+e^- \to (Z\,Z') \to t\bar{t}h$ process.

## 1. Introduction

A simple extension of the Standard Model (SM) is analysed by adding a complex singlet to the scalar sector, the $U(1)_{\rm B-L}$ model [1−5] with an extra $U(1)$ local gauge symmetry [6], where B(L) is baryon (lepton) number. The spontaneous breaking of $U(1)_{\rm B-L}$, incorporates a complex scalar field $\chi$ in addition to the doublet $\Phi$ of Higgs fields. This symmetry plays an important role in various physics scenarios beyond the SM. First, the gauge $U(1)_{\rm B-L}$ symmetry group is contained in the Grand Unification Theory (GUT) described by a $SO(10)$ group [1]. Second, the scale of the B−L symmetry breaking is related to the mass scale of the heavy right-handed Majorana neutrino mass terms and provides the well-known see-saw mechanism [7−11] to explain light left-handed neutrino mass. Third, the B–L symmetry and the scale of its breaking are tightly connected to the baryogenesis mechanism through leptogenesis [12]. The crucial test of the model is the detection of the new heavy neutral $Z'$ gauge boson and the new heavy Higgs boson $H$. The analysis of precision electroweak measurements indicates that the $Z'$ boson should be heavier than about 1.2 TeV. Searches for both of these particles predicted by the B–L model are presently being conducted at the LHC, Therefore, another Higgs

factory besides the LHC, such as the ILC (International Linear Collider) and CLIC (Compact Linear Collider), could precisely determine the properties of the Higgs bosons $h$ and $H$.
The Higgs-strahlung process $e^+e^- \rightarrow Zh$ [13-17] is one of the main production mechanisms of the Higgs boson in $e^+e^-$ linear colliders. After the discovery of the Higgs boson, detailed experimental and theoretical studies are necessary for checking its properties [18-21]. It is possible to search for the Higgs boson in the framework of the B–L model; however, the existence of a new gauge boson could also provide new Higgs particle production mechanisms that could prove its non-standard origin.

We examine the $e^+e^- \rightarrow Zh$ and $ZH$ production modes, and we complement our previous works with the addition of the $e^+e^- \rightarrow (Z\,Z') \rightarrow ht\bar{t}$ process, including the possibility of $Z'$ mediation, which could be resonant, as we allow for $Z/Z'$ mixing. We focus on the B–L model [22-23] due to its relatively simple theoretical structure and background free environment. The crucial test of the model is the detection of the new heavy neutral $Z'$ gauge boson and the new heavy Higgs boson $H$.

## 2. The $U(1)_{B-L}$ model

The solid evidence for the non-vanishing neutrino masses has been confirmed by various neutrino oscillation phenomena [32-33] and indicates the evidence of new physics beyond the SM. The most attractive idea to naturally explain the small neutrino masses is the see-saw mechanism [8-10], in which heavy right-handed (RH) neutrinos singlets under the SM gauge group are introduced.

We consider a $SU(3)_C \times SU(2)_L \times U(1)_Y \times U(1)_{B-L}$ model, which is an elegant and simple extension of the SM [4, 5, 7, 22], where $U(1)_{B-L}$, represents the additional gauge symmetry. The gauge invariant Lagrangian of this model is given by:

$$\mathcal{L} = \mathcal{L}_s + \mathcal{L}_{YM} + \mathcal{L}_f + \mathcal{L}_Y \tag{1}$$

where $\mathcal{L}_s$, $\mathcal{L}_{YM}$ , $\mathcal{L}_f$ , $\mathcal{L}_Y$ are the scalar, Yang-Mills, fermion and Yukawa sector, respectively.

The Lagrangian for the gauge and scalar sectors is given by [4, 24]:

$$\mathcal{L}_g = -\frac{1}{4}W^a_{\mu\nu}W^{a\mu\nu} - \frac{1}{4}B_{\mu\nu}B^{\mu\nu} - \frac{1}{4}B'_{\mu\nu}B'^{\mu\nu} \tag{2}$$

where we see the corresponding field strength tensors for $SU(2)_L \times U(1)_Y \times U(1)_{B-L}$ respectively.

$$\mathcal{L}_s = (D^\mu\Phi)^\dagger(D_\mu\Phi) + (D^\mu\chi)^\dagger(D_\mu\chi) - V(\Phi,\chi) \tag{3}$$

here we can clearly see the added $\chi$ field of our model, and where our new potential term is [25]:

$$V(\Phi,\chi) = m^2(\Phi^\dagger\Phi) + \mu^2|\chi|^2 + \lambda_1(\Phi^\dagger\Phi)^2 + \lambda_2|\chi|^4 + \lambda_3(\Phi^\dagger\Phi)|\chi|^2 \tag{4}$$

with $\Phi$ and $\chi$ as the complex scalar Higgs doublet and singlet fields, respectively. The covariant derivative is given by [25-27]:

$$D^\mu = \partial_\mu + ig_s\,t^a G^\alpha_\mu + i\left[gT_a W^a_\mu + g_1 Y B_\mu + (\tilde{g}Y + g'_1 Y_{B-L})B'_\mu\right] \tag{5}$$

where $g_s$, $g$, $g_1$ and $g'_1$ are the $SU(3)_C \times SU(2)_L \times U(1)_Y \times U(1)_{B-L}$ couplings with $t^a$, $T_a$, $Y$ and $Y_{B-L}$ being their respective group generators. The mixing between the two Abelian groups is described by the new coupling $\tilde{g}$, which is related to the other gauge couplings $g_1$ and $g'_1$, the "pure" or "minimal" B–L model is defined by the condition $\tilde{g} = 0$, that implies no mixing between the B–L $Z'$ and SM $Z$ bosons. In this article we consider the case $\tilde{g} \neq 0$, meaning that we do consider a mixing defined by the angle $\theta_{B-L}$The electromagnetic charges of the fields are the same as those of the SM and the new hypercharges $Y_{B-L}$ are 1/3 for quarks, -1 for leptons, 0 and 2 for the $\Phi$ and $\chi$ fields respectively [4-5, 25-27], to preserve the gauge invariance of the model. The doublet $\Phi$ and singlet $\chi$ scalar fields before and after spontaneous symmetry breaking are given by:

$$\Phi = \begin{pmatrix} G^\pm \\ \frac{v+\phi^0+iG_Z}{\sqrt{2}} \end{pmatrix} \rightarrow \begin{pmatrix} 0 \\ \frac{v+\phi^0}{\sqrt{2}} \end{pmatrix} \qquad \chi = \left(\frac{v'+\phi'^0+iz'}{\sqrt{2}}\right) \rightarrow \left(\frac{v'+\phi'^0}{\sqrt{2}}\right) \tag{6}$$

where $v'$ is the B−L symmetry breaking scale constrained by the electroweak precision measurements data whose value is assumed to be at least of the order of TeV. After minimizing the potential around these values we see that the mass eigenstates are linear combinations of $\Phi^0$ and $\Phi'^0$ and written:

$$\begin{pmatrix} h \\ H \end{pmatrix} = \begin{pmatrix} \cos\alpha & -\sin\alpha \\ \sin\alpha & \cos\alpha \end{pmatrix} \begin{pmatrix} \phi^0 \\ \phi'^0 \end{pmatrix} \qquad (7)$$

$h$ is the SM Higgs boson, $H$ is an extra Higgs boson and the scalar mixing angle $\alpha$ $(-\pi/2 \leq \alpha \leq \pi/2)$.

The extension we are studying is in the Abelian sector of the SM gauge group, so that the charged gauge bosons $W^{\pm}$ will have masses given by their SM expressions related to the $SU(2)_L$ factor only, the other gauge boson masses are not so simple to identify because of mixing. In fact, analogously to the SM, the fields of definite mass are linear combinations of $W_\mu^3$, $B^\mu$, and $B'^\mu$, the relation between the neutral gauge bosons and the corresponding mass eigenstates is given by [22, 23, 26, 27];

$$\begin{pmatrix} B^\mu \\ W^{3\mu} \\ B'^\mu \end{pmatrix} = \begin{pmatrix} \cos\theta_W & -\sin\theta_W\cos\theta_{B-L} & \sin\theta_W\sin\theta_{B-L} \\ \sin\theta_W & \cos\theta_W\cos\theta_{B-L} & -\cos\theta_W\sin\theta_{B-L} \\ 0 & \sin\theta_{B-L} & \cos\theta_{B-L} \end{pmatrix} \begin{pmatrix} A^\mu \\ Z^\mu \\ Z'^\mu \end{pmatrix} \text{ with } \theta_{BL}\ (-\pi/4 \leq \theta_{BL} \leq \pi/4) \qquad (8)$$

## 3. The Higgs-strahlung process $e^+e^- \to (Z, Z') \to Zh$, $e^+e^- \to (Z, Z') \to ZH$ and $e^+e^- \to (Z, Z') \to t\bar{t}h$ in the B - L model

The corresponding Feynman diagrams are shown in figure 1 and we calculate the Higgs production cross section $\sigma$ in the context of the B − L model considering both, a $Z$ and a $Z'$ mediator.

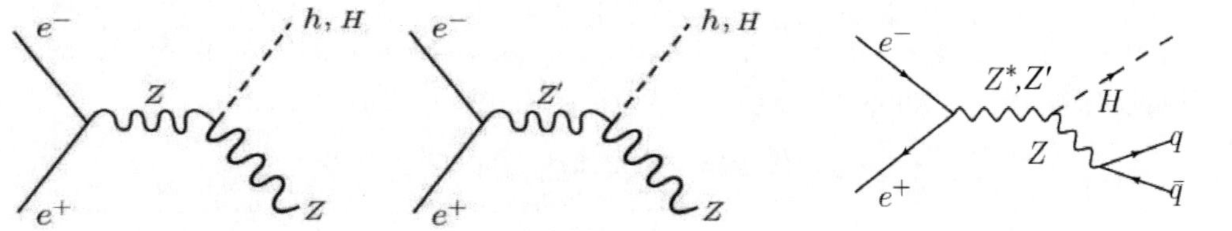


Figure 1
Feynman diagrams for the processes $e^+e^- \to (Z, Z') \to Zh$, $e^+e^- \to (Z, Z') \to ZH$ and $e^+e^- \to (Z, Z') \to t\bar{t}h$

Higgs-strahlung cross sections $\sigma$ for the production of the SM like Higgs boson $h$ [34];

$$\sigma_Z(e^+e^- \to Zh) = \frac{G_F^2 M_Z^4 \cos^2\alpha\left((g_v^e)^2 + (g_a^e)^2\right) S\sqrt{\lambda}}{24\pi\left(\left(s - M_Z^2\right)^2 + M_Z^2\Gamma_Z^2\right)}\left(\lambda + \frac{12M_Z^2}{s}\right) \qquad (9)$$

$$\sigma_{(Z')}(e^+e^- \to Zh) = \frac{G_F^2 M_Z^6}{24\pi}\left((g_v'^e)^2 + (g_a'^e)^2\right)\frac{s\sqrt{\lambda}\left[\lambda + 12M_{Z'}^2/s\right]}{M_{Z'}^2\left[\left[s - M_{Z'}^2\right]^2 + M_{Z'}^2\Gamma_{Z'}^2\right]}\ \left[f(\theta')\cos\alpha + g(\theta')\text{sen}\alpha\right]^2 \qquad (10)$$

$$\sigma_{(Z,Z')} = \frac{G_F^2 M_Z^6 \cos\alpha}{6\pi}(g_v'^e g_v^e + g_a'^e g_a^e)s\sqrt{\lambda}\left[\frac{1}{M_Z^2}\left(\lambda + 12\frac{M_Z^2}{s}\right) + \frac{1}{M_{Z'}^2}(\lambda + 6\,(M_Z^2 - M_{Z'}^2)/s)\right.$$
$$\left. + \frac{s\lambda}{8M_Z^2 M_{Z'}^2}\left(\lambda - 12\frac{M_Z^2}{s}\right)\right]\frac{\left[(s - M_Z^2)(s - M_{Z'}^2) + M_Z M_{Z'}\Gamma_Z\Gamma_{Z'}\right]}{\left[(s - M_Z^2)^2 + M_Z^2\Gamma_Z^2\right]\left[\left(s - M_{Z'}^2\right)^2 + M_{Z'}^2\Gamma_{Z'}^2\right]}\left[f(\theta')\cos\alpha + g(\theta')\text{sen}\alpha\right] \qquad (11)$$

where

$$\lambda(M_Z^2, s, M_h^2) = \left(1 - \frac{M_Z^2}{s} - \frac{M_h^2}{s}\right)^2 - 4\frac{M_Z^2 M_h^2}{s^2} \qquad (12)$$

is the usual two particle phase space function.

The expression given in equation (9) corresponds to the cross section with the exchange of the $Z$ boson, while equations (10) and (11) come from the contributions of the B–L model and of the interference, respectively.

Finally we obtain the corresponding cross section for the $e^+e^- \to (Z, Z') \to t\bar{t}h$ process, starting with the transition amplitude in the B-L model.

First we consider the neutral SM boson $Z$ as a mediator, and then the heavy $Z'$ boson of the B-L model.

$$\mathcal{M}_Z = \frac{-ig}{\cos\theta_w}\bar{v}(p_1)\frac{1}{2}\gamma^\mu(g_v^e - g_a^e\gamma_5)u(p_2)\frac{(-g_{\mu\nu}+p_\mu p_\nu/M_Z^2)}{p^2-M_Z^2-i\Gamma_Z^2}\left[\frac{2M_Z^2\cos\alpha}{v}\right]\epsilon_\lambda^{*\nu}(z)\frac{-ig}{\cos\theta_w}\bar{v}(t)\frac{1}{2}\gamma^\mu(g_v^t - g_a^t\gamma_5)u(t) \tag{13}$$

$$\mathcal{M}_{Z'} = \frac{-ig}{\cos\theta_w}\bar{v}(p_1)\gamma^\mu\frac{1}{2}(g_v^e - g_a^e\gamma_5)u(p_2)\left(-g_{\mu\nu}+\frac{p_\mu p_\nu}{M_Z'^2}\right)\left(\frac{1}{p_Z^2-M_Z'^2-i\Gamma_Z^2}\right)\frac{2M_Z'^2}{v}[f(\theta')\cos\alpha + g(\theta')\mathrm{sen}\alpha]\bar{v}(t)\frac{-ig}{\cos\theta_w}\gamma^\alpha\frac{1}{2}(g_v'^t - g_a'^t\gamma_5)u(t) \tag{14}$$

and the cross section $\sigma$ is given by;

$$\frac{d\sigma(e^+e^-\to f\bar{f}H)}{dx_1dx_2} = \frac{N_c\sigma_0}{4\pi}\left\{\left[Q_e^2Q_f^2 + \frac{2Q_eQ_fv_ev_f}{1-z} + \frac{(v_e^2+a_e^2)(v_f^2+a_f^2)}{(1-z)^2}\right]G_1 + \frac{v_e^2+a_e^2}{(1-z)^2}\left[a_f^2\sum_{i=2}^6 G_i + v_f^2(G_4+G_6)\right] + \frac{Q_eQ_fv_ev_f}{1-z}G_6\right\} \tag{15}$$

$\sigma_0 = 4\pi\alpha^2/3s$ is the standard normalization cross section, with $\alpha = 1/127$ for the value of the running electromagnetic coupling constant, $x_1, x_2$, are the reduced energies of the fermions, $2E_{f,\bar{f}}/\sqrt{s}$, $Q_f$ is the electric charge, $N_c$ the number of quark colors, $a_f$ and $v_f$ are the axial and vectorial coupling constants or $Z$ charges of the fermion $f$, respectively, and normalized defined as [35]

$a_f = \frac{2I_{3L}^f}{4c_Ws_W}$; $v_f = \frac{2I_{3L}^f - 4Qs_W^2}{4c_Ws_W}$, where, $I_{3L}^f = \pm\frac{1}{2}$ is the weak isospin of the left handed fermions and $s_W^2 = 1 - c_W^2 = \sin^2\theta_w = 0.23$

The coefficients $G_1$ and $G_2$ describe the Higgs radiation off the fermion $f$, the other terms account for the emission of the Higgs particle from the $Z$-boson line ($G_3$, $G_4$), and the interference terms ($G_5$, $G_6$) between the radiation amplitudes off the fermion and the $Z$-boson lines, which we are not considering here.

$$G_1 = \frac{g_{ffH}^2}{x_{12}}\left[x_H^2 - h\left(\frac{x_H^2}{x_{12}} + 2(x_H - 1 - h)\right) + 2f\left(4(x_H - h) + \frac{x_H^2}{x_{12}}(4f - h + 2)\right)\right] \tag{16}$$

$$G_2 = -\frac{2g_{ffH}^2}{x_{12}}\left[x_{12}(1+x_H) - h(x_{12} + 2x_H + 8f - 2h) + 3fx_H\left(\frac{x_H}{3} + 4 + \frac{x_H}{x_{12}}(4f - h)\right)\right] \tag{17}$$

$$G_3 = 2\frac{g_{ZZH}^2}{x_Z^2}\left[f(4h - x_H^2 - 12z) + \frac{f}{z}(4h - x_H^2)(x_H - 1 - h + z)\right] \tag{18}$$

$$G_4 = 2\frac{g_{ZZH}^2}{x_Z^2}z[h + x_{12} + 2(1 - x_H) + 4f] \tag{19}$$

$$G_5 = -\frac{g_{ffH}\,g_{ZZH}}{x_{12}x_Z}4x_H\frac{m_f}{m_Z}\left[(x_{12} - h)(x_H - 1 - h) + f(12z - 4h + x_H^2) - 3z\left(h - 2\frac{x_{12}}{x_H}\right)\right] \tag{20}$$

$$G_6 = -\frac{g_{ffH}\,g_{ZZH}}{x_{12}x_Z}4z\frac{m_f}{m_Z}[x_H(h - 4f - 2) - 2x_{12} + x_H^2] \tag{21}$$

where $x_{12} = (1 - x_1)(1 - x_2)$ and $x_H = 2E_H/\sqrt{s} = 2 - x_1 - x_2$ is the reduced energy of the Higgs boson.

## 4. Results and conclusions

We evaluate the total cross section $\sigma$ of the process $e^+e^- \to (Z, Z') \to Zh$ in the B-L model using these values; [31, 34]; $\sin^2\theta_W = 0.23126 \pm 0.00022$ , $m_\tau = 1776.82 \pm 0.16$ MeV, $m_b = 4.18 \pm 0.06$ GeV, $m_t = 172.44 \pm 0.13$ GeV, $M_W = 80.389 \pm 0.023$ GeV, $M_Z = 91.1876 \pm 0.0021$ GeV, $\Gamma_Z = 2.4952 \pm 0.0023$ GeV, $M_h = 125.09 \pm 0.4$ GeV. If we consider the recent limit from $\frac{M_{Z'}}{g'_1} \geq 6.9$ TeV [28-30], it is possible to obtain a direct bound on the B–L breaking scale $v'$, and take $v' = 3.45$ TeV and $\alpha = \pi/9$ in our numerical analysis. We will assume $\sqrt{s}$, $M_{Z'}$, $g'_1$, $\theta_{B\text{-}L}$ as free parameters.

In figure 2, we show the cross section $\sigma\,(e^+e^- \to Zh)$, for different contributions as a function of the center-of-mass energy $\sqrt{s}$ for $\theta_{B\text{-}L} = 10^{-3}$ and $g'_1 = 0.290$: the solid line corresponds to the SM and the dashed line corresponds to $\sigma_Z$ where the $U(1)_{B-L}$ model contributes to the couplings $g^f_V$ and $g^f_A$ of the SM gauge boson $Z$ to electrons. The dot-dashed line corresponds to $\sigma_{z'}$, which is only the B-L contribution, the dot dot-dashed line corresponds to the interference $\sigma_{z,z'}$. Finally, the dotted line corresponds to the total cross section, $\sigma_{Tot.}$. We can see that the cross section corresponding to $\sigma_Z(e^+e^- \to Zh)$ decreases for large $\sqrt{s}$, whereas the cross section of the B-L model and the total cross section, there is an increase for large values of the center-of-mass energy, reaching its maximum value at the resonance of $Z'$, $\sqrt{s} = 2000$ GeV.

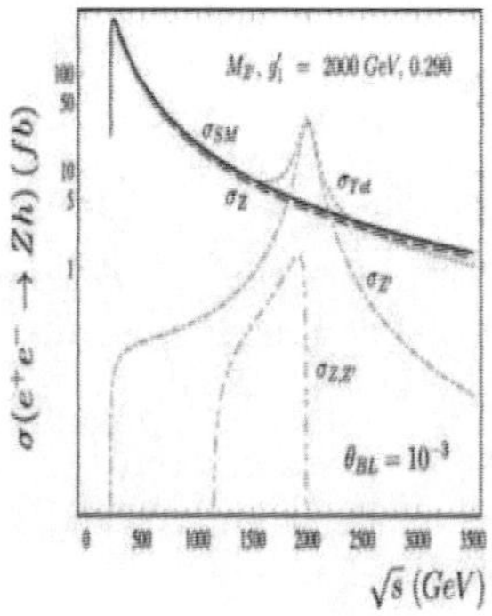


Figure 2

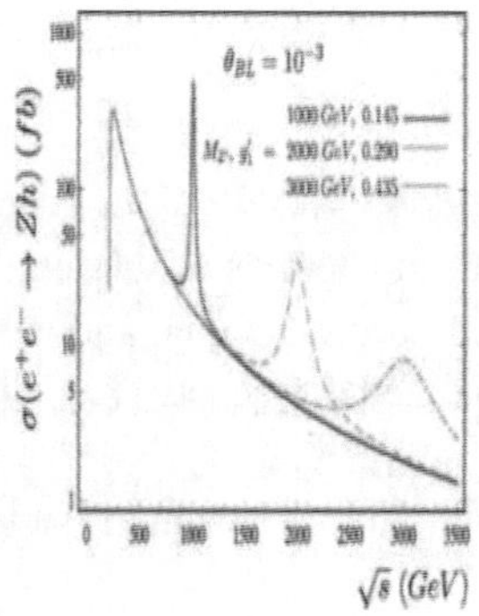


Figure 3

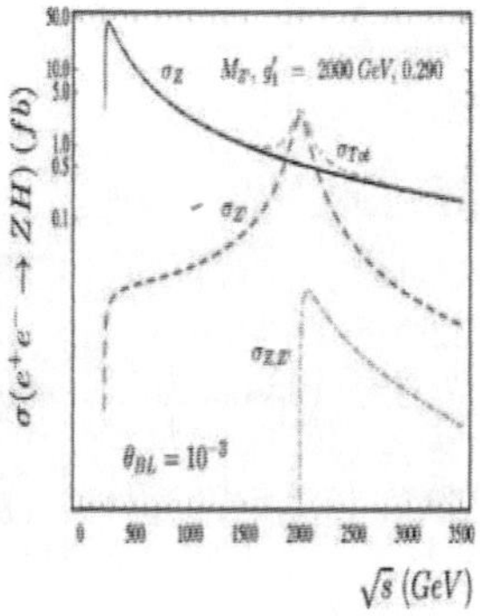


Figure 4

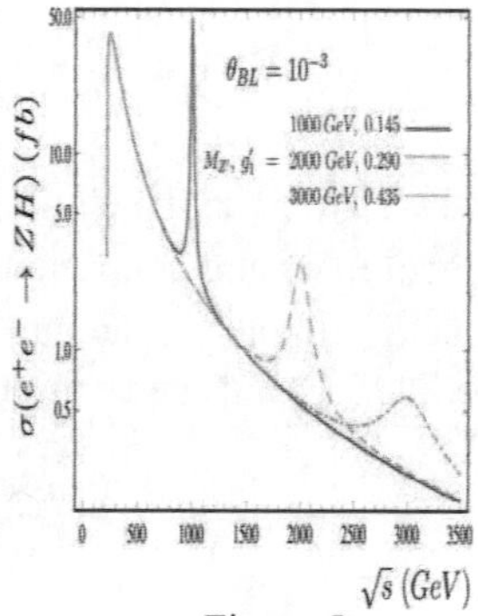


Figure 5

Total cross sections of the processes, $e^+e^- \to Zh$ (2, 3), and $e^+e^- \to ZH$ (4, 5) as a function of the center of mass energies $\sqrt{s}$, with $\theta_{B\text{-}L} = 10^{-3}$ for different values of $M_{Z'}$ and $g'_1$

We plot the total cross section of the reaction $e^+e^- \to Zh$ in figure 3 as a function of the center-of-mass energy $\sqrt{s}$, for values of the heavy gauge boson mass $M_{Z'} = 1000$, 2000, 3000 GeV and $g'_1 = 0.145$, 0.290, 0.435, respectively. $M_{Z'}$ and $g'_1$ maintain the previous relationship. Here we see that the cross section is sensitive to the free parameters and also that the height of the resonance peaks for the boson $Z'$ changes depending on the value of $\sqrt{s} = M_{Z'}^2$. In addition, the resonances are broader for larger $g'_1$ values, as the total width of the $Z'$ boson increases with $g'_1$.

The total cross section for the production processes $e^+e^- \to ZH$ as a function of the collision energy for $M_h = 125$ GeV, $M_H = 800$ GeV, $M_{\nu_R} = 300$ GeV, $M_{Z'} = 2000$ GeV and $g'_1 = 0.290$ is shown in figure 4. Here the curves are for $\sigma_Z(e^+e^- \to ZH)$ (solid line), $\sigma_{Z'}(e^+e^- \to ZH)$ (dashed line), $\sigma_{Z,Z'}(e^+e^- \to ZH)$ (dot-dashed line), and the dot dot-dashed line corresponds to the total cross section of the process $\sigma_{Tot}(e^+e^- \to ZH)$.

To see the effects of $\theta_{B\text{-}L}$, $g'_1$, $M_{Z'}$, the free parameters of the B-L model, we plot the total cross section of the process $e^+e^- \to ZH$ in figure 5 as a function of the center-of mass energy $\sqrt{s}$ for the values of the heavy gauge boson mass of $M_{Z'} = 1000$ GeV with $g'_1 = 0.145$, $M_{Z'} = 2000$ GeV with $g'_1 = 0.290$ and $M_{Z'} = 3000$ GeV with $g'_1 = 0.435$, preserving the relationship between $M_{Z'}$ and $g'_1$. In this figure we observed that for $\sqrt{s} = M_{Z'}$ the resonant effect dominates, the cross section is sensitive to the free parameters. We also observe that the height of the resonance peaks for the $Z'$ boson change depending on the value of $\sqrt{s} = M^2_{Z'}$, and in addition, we see that the resonances are broader for larger $g'_1$ values, as the total width of the $Z'$ boson increases with $g'_1$.

We see that the expected $Z_h$ and $Z_H$ events can reach a number of $\mathcal{O}$ ($10^6$ and $10^5$), respectively, which is a very optimistic scenario and it would be possible to perform precision measurements for both Higgs bosons $h$ and $H$, for the $Z'$ heavy gauge boson, as well as for the parameters of the model $\theta_{B-L}$, $g'_1$ and $\alpha$. In addition, the SM expression for the cross section of the reaction $e^+e^- \to Zh$ can be obtained in the decoupling limit, when $\theta_{B-L}$, $g'_1 = 0$ and $\alpha = 0$. Our study complements other studies on the B−L model and on the Higgs-strahlung processes; $e^+e^- \to (Z, Z') \to Zh$ $(ZH)$.

## References

[1] Buchmuller W, Greub C and Minkowski P 1991 *Phys. Lett.* B **267** 395
[2] Marshak R and Mohapatra R N 1980 *Phys. Lett.* B **91** 222
[3] Mohapatra R N and Marshak R 1980 *Phys. Rev. Lett.* **44** 1316
[4] Khalil S 2008 *J. Phys. G: Nucl. Part. Phys*. G **35** 055001
[5] Eman W and Khalil 2007 S *Eur. Phys. J. C* **52** 625
[6] Carlson E D 1987 *Nucl. Phys* B **286** 378
[7] Mohapatra Rabindra N and Senjanovic G 1980 *Phys. Rev. Lett.* **44** 912
[8] Minkowski P 1977 *Phys. Lett.* B **67** 421
[9] Van Nieuwenhuizen P and Freedman D Z 1979 *Proc. Supergravity Conf.* (Stony Brook University) ed P Van Nieuwenhuizen and D Z Freedman (Amsterdam, Netherlands: North Holland Publishing Co.) p 341
[10] Yanagida T 1979 *Proc. Workshop on the Baryon Number of the Universe and Unified Theories* (Tsukuba, Japan, 13–14 February) p 95
[11] Gell-Mann M, Ramond P and Slansky R 1979 *Proc. Supergravity Conf.* (Stony Brook University, New York) ed P Van Nieuwenhuizen and D Z Freedman (Amsterdam, Netherlands: North Holland Publishing Co.) pp 315–21
[12] Fukugita M and Yanagida T 2003 *Physics of Neutrinos and Applications to Astrophysics* (Berlin: Springer)
[13] Ellis J, Gaillard M K and Nanopoulos D V 1976 *Nucl. Phys*. B **106** 292
[14] Ioffe B L and Khoze V A 1978 Sov. *J. Part. Nucl.* **9** 50
[15] Lee B W, Quigg C and Thacker H B 1977 *Phys. Rev.* D **16** 1519
[16] Bjorken J D 1976 *Proc. Summer Institute on Particle Physics* SLAC Report 198
[17] Barger V D *et al* 1994 *Phys. Rev.* D **49** 79
[18] Ellis J 2013 arXiv:1312.5672
[19] Dawson S *et al* 2013 arXiv:1310.8361
[20] Klute M *et al* 2013 *Europhys. Lett.* **101** 51001
[21] Behnke T *et al* 2013 arXiv:1306.6327 [physics.acc-ph]
[22] Basso L *et al* 2009 *Phys. Rev.* D **80** 055030
[23] Basso L *et al* 2009 *J. High Energy Phys*. JHEP10(2009)006
[24] Basso L 2011 arXiv:1106.4462 [hep-hp]
[25] Basso L, Moretti S and Pruna G M 2010 *Phys. Rev.* D **82** 055018
[26] Basso L *et al* 2011 *Eur. Phys. J.* C **71** 1613
[27] Basso L, Moretti S and Pruna G M 2012 *J. Phys. G: Nucl. Part. Phys*. G **39** 025004
[28] Carena M, Daleo A, Dobrescu B A and Tait T M 2004 *Phys. Rev.* D **70** 093009
[29] Heeck J 2014 *Phys. Lett.* B **739** 256
[30] Cacciapaglia G, Csaki C, Marandella G and Strumia A 2006 *Phys. Rev.* D **74** 033011
[31] Olive K A *et al* (Particle Data Group) 2014 *Chin. Phys.* C **38** 090001
[32] Fogli G L, Lisi E, Marrone A, and Palazzo A 2006 *Prog. Part. Nucl. Phys*. **57** 742
[33] Fogli G L *et al* 2007 *Phys. Rev.* D **75** 053001
[34] Ramírez-Sánchez F *et al* 2016 *J. Phys. G: Nucl. Part. Phys.* **43** 095003
[35] Djouadi A, Kalinowski J and Zerwas P M 1992 *Z. Phys. C Particles and Fields* **54**, 255-262

IOP Conf. Series: Journal of Physics: Conf. Series **1078** (2018) 012019 doi:10.1088/1742-6596/1078/1/012019

# Splitting of single-particle levels in clusters potentials

**A.H. Santana-Valdés* and R. Bijker**

Instituto de Ciencias Nucleares, Universidad Nacional Autónoma de México, A.P. 70-543, 04510 México, D.F., México

E-mail: *adrian.santana@correo.nucleares.unam.mx

**Abstract.** In analogy with the Nilsson model, we calculate the splitting of spherical single-particle levels in a deformed field, but for cluster potentials. We study applications to alpha-cluster nuclei with two, three and four alpha particles, in which the deformation corresponds to the relative distance between the alpha particles. The splitting of the single-particle levels is studied for the cases of a dumbbell, equilateral triangle and a regular tetrahedron. The observed patterns may be used to gain insight into how the single-particle levels evolve with deformation.

## 1. Introduction

Energy level splitting is a very well-known phenomenon in physics, which occurs whenever a set of degenerate levels is split in an external field. One of the best known examples is the Zeeman effect in which rotational states are split in an external magnetic field according to their magnetic substates (Figure 1) [1].

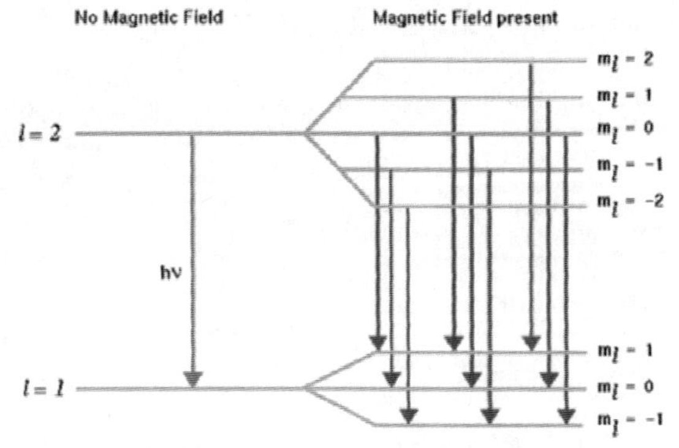


**Figure 1.** Zeeman effect: splitting of levels due to an external magnetic field.

Another example is the Nilsson model of deformed shell model where the single-particle levels are split in a deformed quadrupole field [2]. In this case the energy levels are not only split but also strongly mixed by the quadrupole field. The amount of mixing depends on the strength of the deformed field (Figure 2). Level crossing and repulsion occur due to this deformation. In the case of the Zeeman effect, the levels are split but not mixed, and can therefore be labeled by the same quantum numbers as before the splitting, i.e the angular momentum $l$ and its projection $m_l$. In the case of the Nilsson model, the levels are strongly mixed and are labeled by the quantum numbers in the intrinsic or body-fixed frame $[Nn_z\Lambda\Omega]^\pi$ where $\Omega$ denotes the

IOP Conf. Series: Journal of Physics: Conf. Series **1078** (2018) 012019 doi:10.1088/1742-6596/1078/1/012019

projection of the single-particle angular momentum on the symmetry axis (z-axis), $\pi$ is the parity, $N$ the principal quantum number, $n_z$ the number of nodes in the wave function in the z-direction and $\Lambda$ the projection of the orbital angular momentum on the symmetry axis.

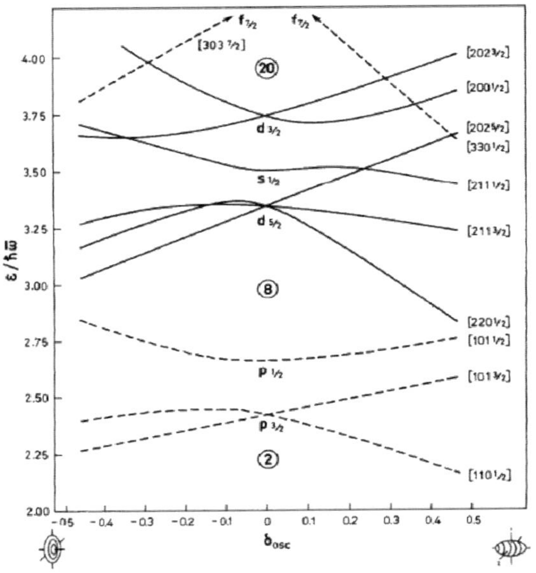


**Figure 2.** Splitting of single-particle levels as a function of the quadrupole deformation in the deformed shell model. Levels with even and odd parity are drawn with solid and dashed lines respectively (taken from [1]).

In our work, we use a deformed field, that of cluster potentials, where the clusters are alpha particles. The deformation is taken to be the relative distance to the center of mass of the whole $k\alpha + x$ nucleon structure, where is $x$ a nucleon. The splitting of the single-particle levels is studied as a function of the relative distance for the cases of a dumbbell, equilateral triangle and a regular tetrahedron, where a symmetry factor is crucial for the level splitting and configuration mixing. Relabeling is also necessary in this case specially for the cases of an equilateral triangle and regular tetrahedron. The main goal for this study is to obtain patterns that gain insight into how the single-particle levels evolve with deformation. That is done by studying the mixing between states, how these are related with the geometrical configuration and what kind of symmetry is present.

## 2. Cluster Shell Model

We start by reviewing the Cluster Shell Model [3]. In this model nuclei with $Z = N = 2k$ are treated as a cluster of $k$ $\alpha$-particles whose matter and charge density are given by a gaussian form

$$\rho(\vec{r}) = \left(\frac{\alpha}{\pi}\right)^{\frac{3}{2}} e^{-\alpha\left(r^2+\beta^2\right)} 4\pi \sum_{\lambda\mu} i_\lambda(2\alpha\beta r)\, Y_{\lambda\mu}(\theta,\phi) \sum_{i=1}^{k} Y^*_{\lambda\mu}(\theta_i,\phi_i)\,. \tag{1}$$

Here $\vec{r}_i = (\beta, \theta_i, \phi_i)$ where $\beta$ denotes the relative distance of the $\alpha$ particles to the center of mass, and $\theta_i$ and $\phi_i$ are the angles. An important factor is the cluster factor, $\sum_{i=1}^{k} Y^*_{\lambda\mu}(\theta_i,\phi_i)$, which contains the information about the geometrical configuration of the alpha particles.
The potential is obtained by convoluting the density, eq. (1), with a Volkov potential [4] to obtain

$$V(\vec{r}) = -V_0 \sum_{\lambda\mu} 4\pi e^{-\alpha\left(r^2+\beta^2\right)} i_\lambda(2\alpha\beta r)\, Y_{\lambda\mu}(\theta,\phi) \sum_{i=1}^{k} Y^*_{\lambda\mu}(\theta_i,\phi_i)\,. \tag{2}$$

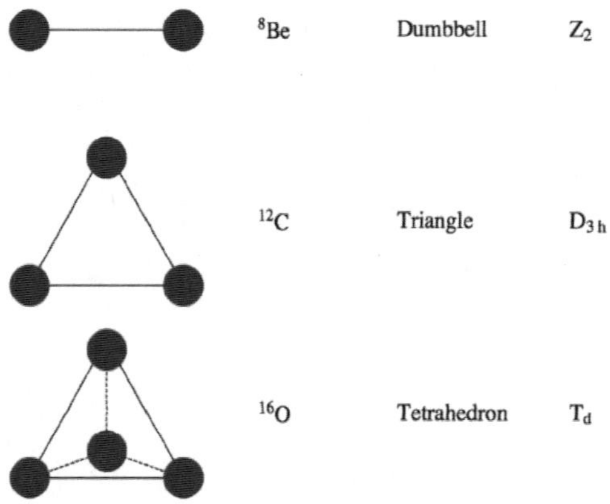


**Figure 3.** Geometrical configuration for $k = 2, 3, 4$ clusters (taken from [3]).

The spin-orbit interaction is taken as

$$V_{so}(\vec{r}) = V_{0,so}\frac{1}{2}\left[\frac{1}{r}\frac{\partial V(\vec{r})}{\partial r}\left(\vec{s}\cdot\vec{l}\right) + \left(\vec{s}\cdot\vec{l}\right)\frac{1}{r}\frac{\partial V(\vec{r})}{\partial r}\right]. \tag{3}$$

Finally, the coulomb potential is found by convoluting the carge density $\frac{Ze^2}{k}\rho(\vec{r})$ with Green's function giving

$$V_C(\vec{r}) = \frac{Ze^2}{k}\left(\frac{\alpha}{\pi}\right)^{\frac{3}{2}}\sum_{\lambda\mu}\frac{4\pi}{2\lambda+1}Y_{\lambda\mu}(\theta,\phi)\sum_{i=1}^{k}Y^*_{\lambda\mu}(\theta_i,\phi_i)\times \tag{4}$$

$$\times\left[\frac{1}{r^{\lambda+1}}\int_0^r f_\lambda(r')\,r'^{\lambda+2}dr' + r^\lambda\int_r^\infty f_\lambda(r')\frac{1}{r'^{\lambda-1}}dr'\right]. \tag{5}$$

We then obtain the single-particle energy levels and intrinsic states as a function of $\beta$ for each configuration ($k = 2, 3, 4$) of fig. 3 by solving the single-particle Schrödinger equation

$$H = \frac{\vec{p}^2}{2m} + V(\vec{r}) + V_{so}(\vec{r}) + V_C(\vec{r}). \tag{6}$$

In the case of neutrons $V_C(\vec{r}) = 0$.

## 3. Results

The Hamiltonian of eq. (6) is diagonalized in the harmonic oscillator basis. The correlation diagrams are shown as a function of $\beta$ in Figures 4 and 5. The states for the spherical case correspond to $\beta = 0$. For the two-body cluster configuration the levels with projection $\pm\Omega$ are degenerate. As a consequence, a single-particle level $(l, 1/2)\,j$ is split into a series of doublets with $\Omega = 1/2,\ 3/2,\ \ldots,\ j$ and parity $P = (-)^l$. For the triangular configuration neither the angular momentum nor parity is conserved. The spherical single-particle levels are split into a series of doublets characterized by $E'_2$ and $E'_1$, the double valued representations of $D_3$, the rotational subgroup of $D_{3h}$. For the tetrahedral configuration, the single-particle levels are split into a

**Table 1.** Resolution of single-particle levels into the double-valued representations of the corresponding point group. Notation from Hamermesh [5].

| $j$ | Dumbbell | Triangle | Tetrahedron |
|---|---|---|---|
| $1/2$ | $(1/2)^P$ | $E'_2$ | $E'$ |
| $3/2$ | $(1/2+3/2)^P$ | $E'_1+E'_2$ | $G'$ |
| $5/2$ | $(1/2+3/2+5/2)^P$ | $E'_1+2E'_2$ | $E'+G'$ |
| $7/2$ | $(1/2+3/2+5/2+7/2)^P$ | $E'_1+3E'_2$ | $2E'+G'$ |

series of doublets and quadruplets characterized by $E'$ and $G'$, respectively, the double valued representations of $T$, the rotational subgroup of $T_d$. The results are summarized in table 1.
The first result that we can see by comparing Figure 4 is that the correlation diagrams for protons and neutrons behave in the same manner. This result shows that it does not matter much whether the particle is a neutron or a proton, there is just a small shift in energy.
In order to adress the issue of mixing between the levels we study the overlap between intrinsic states of different values of deformation, $\langle \chi_\Omega(\beta) | \chi_{\Omega'}(\beta') \rangle$. We can then have the overlap with a fixed value of $\beta'$, namely $\beta' = 0$. In that way we can see the mixing between states as we vary $\beta$. This can be seen in Figure 5, where from comparing the graphs of overlap of the triangle and tetrahedron, there is present more mixing in the case of triangle than the case of tetrahedron. A main reason why this occurs is because the density of states of same symmetry is higher in the configuration of triangle, as a result those levels are then closer energetically to each other and can mix more strongly in comparison to the tetrahedron.

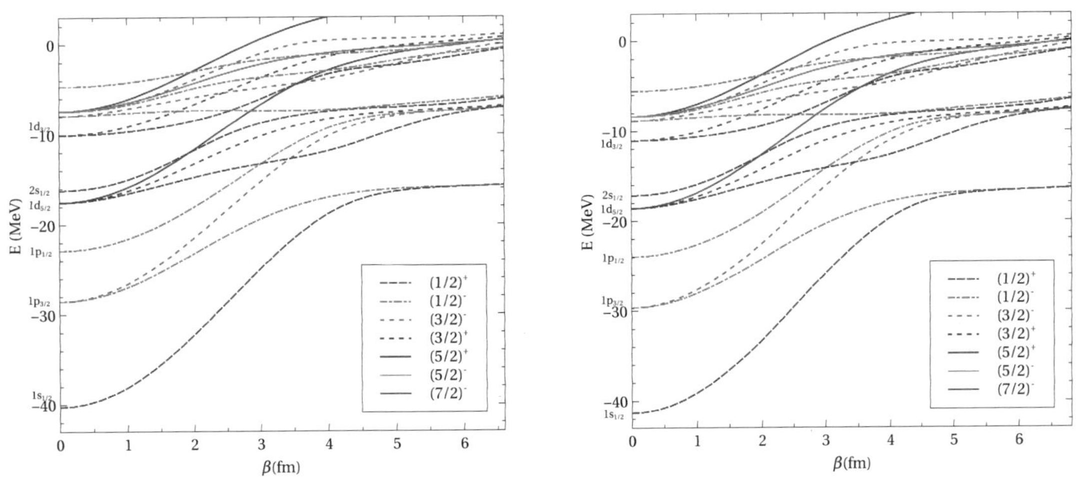


**Figure 4.** Correlation diagrams for the case of dumbbell, obtained from eq. (6) with $V_0 = 32$ MeV, $\alpha = 0.0511$ fm$^{-2}$ and $V_{0,so} = 27.5$ MeV fm$^2$ [3] for a proton (left) and a neutron (right).

For the case of triangle there exist strong repulsion between states, which causes crossings of wave function from one level to another. These cases can be seen in the graph of overlap of the triangle in the region 3.5 to 4.5 fm for the states belonging to the $E'_2$ symmetry. In states

IOP Conf. Series: Journal of Physics: Conf. Series **1078** (2018) 012019 doi:10.1088/1742-6596/1078/1/012019

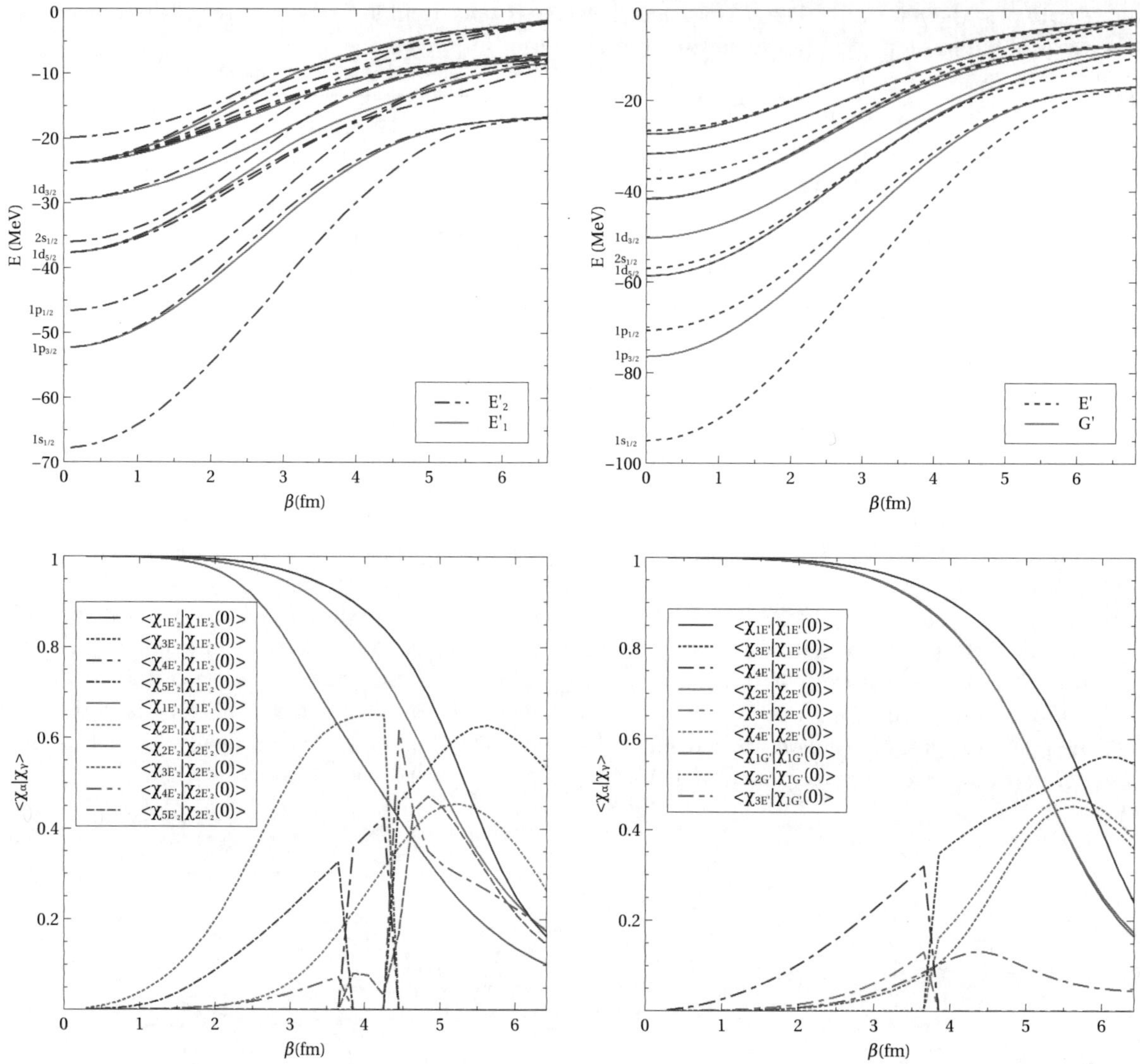


**Figure 5.** Correlation diagrams (top) and overlaps between states of the same symmetry (bottom) for the case of a triangle (left) and a tetrahedron (right), and a single-particle neutron, obtained from eq. (6) with $V_0 = 32$ MeV, $\alpha = 0.0511$ fm$^{-2}$ and $V_{0,so} = 17$ MeV fm$^2$ for the triangle, and $V_{0,so} = 12.1$ MeV fm$^2$ for the tetrahedron [3].

belonging to the $E_1'$ that doesn't happen, and the mixing occurs smoothly with no wave function crossings.

In the case of the tetrahedron, although there occurs a wave function crossing (between 3.5 to 4 fm) this doesn't happens as much as in the case of triangle. There is strong mixing but states are sufficiently separated so that wave function crossings don't occur as often as in the case of triangle.

In both cases the difference between states of the same symmetry group arise from how much energetically distanced they are from other states. A simple example are the ground levels, which in comparison to higher levels present much less mixing. That can be seen by comparing the curves $\langle \chi_\Omega (\beta) | \chi_\Omega (\beta' = 0) \rangle$ where $\Omega$ can be any of the double-valued representations for

triangle and tetrahedron shown in Table 1. The faster those curves decrease the more the state has mixed with others. Again, the mayor difference between the states is seen in the case of the triangle. Also the higher the state of the same group of symmetry the more it mixes.
From both correlation diagrams of Figure 5 it can be seen that the single-particle levels asymptotically tend to a certain degeneracy, which corresponds to the case where the alpha particles are too far apart and thus interact as separated alpha particles. This is something that is not obtained in the Nilsson model, where the degeneracy shown is that of an elongated cigar. As this occurs mixing between states decrease. This is shown in the curves of $\langle \chi_{\Omega}(\beta) | \chi_{\Omega'}(\beta' = 0) \rangle$. Here $\Omega$ and $\Omega'$ are states of the same symmetry, but not the same state. The decrease starts in most cases at 5 fm, hinting that the frontier between the moment where the alphas can still be seen as to be part of the same nucleus and not separate particles is around that value. In the correlation diagrams it is also shown that convergence towards the degeneracy of separate alphas occur after 5 fm.

## 4. Conclusions

From the obtained patterns it has been illustrated how single-particle levels evolve with deformation and which states interact most and why. From the overlap it can also be thought which interactions could be the predominant ones (quadrupole, octupole, etc). This and the interpretation of the behaviour due to the symmetry will be further analysed. All the information from this study can later be used to extend the Algebraic Cluster Model (ACM)[6][7] to odd-cluster nuclei like $^{9}$Be, $^{9}$B, $^{13}$C, $^{13}$N, $^{17}$O and $^{17}$F, by developing the Algebraic Cluster-Fermion Model (ACFM)[8].

## Acknowledgments

This work was supported in part by grant IN109017 from DGAPA-UNAM, Mexico.

## References

[1] Bohr A and Mottelson B R 1998 *Nuclear Structure* vol 2 (World Scientific)
[2] Casten R 2000 *Nuclear Structure from a Simple Perspective* (Oxford University Press)
[3] Della Rocca V , Bijker R and Iachello F 2017 *Nucl. Phys.* A **966** 158-84
[4] Volkov A B 1965 *Nucl. Phys.* **74** 33-58
[5] Hamermesh M 1962 *Group Theory and Its Application to Physical Problems* (Dover Publications)
[6] Bijker R and Iachello F 2000 *Phys. Rev.* C **61** 067305
Bijker R and Iachello F 2002 *Ann. Phys.* (N.Y.) **298** 334
[7] Bijker R and Iachello F 2014 *Phys. Rev. Lett.* **112** 152501
Bijker R and Iachello F 2017 *Nucl. Phys.* A **957** 154
[8] Santana-Valdés A H and Bijker R 2018 Work in progress

IOP Conf. Series: Journal of Physics: Conf. Series **1078** (2018) 012020 doi:10.1088/1742-6596/1078/1/012020

# Investigation of Incomplete Fusion events: Recent Results

**Vijay R Sharma[1,⋆], Mohd. Shuaib[2], Pushpendra P. Singh[3,Ψ], Abhishek Yadav[4], Manoj K. Sharma[5] R. Kumar[4], B. P. Singh[2,†], S. Muralithar[4], R. P. Singh[4], R. K. Bhowmik[4] and R. Prasad[2]**

[1]Departamento del Acelerador, Instituto Nacional de Investigaciones Nucleares, Apartado Postal 18-1027, Caixa Postal 11801, Distrito Federal, Mexico
[2]Accelerator Laboratory, Department of Physics, Aligarh Muslim University, Aligarh 202 002, India
[3]Department of Physics, Indian Institute of Technology, Ropar, PB-140 001, India
[4]NP-Group, Inter-University Accelerator Centre, New Delhi-110 067, India
[5]Physics Department, S. V. College, Aligarh-202 001, India

E-mail: ⋆phy.vijayraj@gmail.com, Ψpps@iitrpr.ac.in and †bpsinghamu@gmail.com

**Abstract.** The dynamics of incomplete fusion processes has been extensively investigated and is found to compete with complete fusion at low incident energies (i.e. $\leq$ 7.0 MeV/A). Recent reports suggest that the probability of incomplete fusion depends on $Q_\alpha$ of the projectile, and found to be originated from the high input angular momentum provided into the system due to non-central collisions. Apart from the well documented existence of low energy incomplete fusion, a strong contradiction on previously established mass-asymmetry systematics has been noticed recently. It has been found that the incomplete fusion fraction increases with entrance channel mass-asymmetry for the individual projectiles. Further, a strong dependence of incomplete fusion on the Coulomb effect ($Z_P Z_T$) has been observed and is found to increase linearly with $Z_P Z_T$, a significantly important approach to understand the breakup fusion reactions. For better insights into the onset and strength of incomplete fusion in terms of various entrance channel parameters and to understand the role of high angular momenta in the onset of incomplete fusion, a quality data have been obtained in a variety of experiments performed at the Inter-University Accelerator Center (IUAC), New Delhi, India. Some of the recent results are presented in this paper, conclusively demonstrating the effect of entrance channel parameters on the onset and strength of incomplete fusion, and the usefulness of incomplete fusion to populate high-spin states in final reaction products.

## 1. Introduction

Fusion reactions between heavy-ions constitutes an important field of research in low energy nuclear physics with stable nuclei and are likely to become more important when radioactive beams of high intensity will be available [1, 2, 3, 4, 5]. At low energies, starting from the Coulomb barrier to 7.0 MeV/A, the study of reaction mechanism concerning the transition between complete fusion (CF) and incomplete fusion (ICF) processes is important in the view of understanding the interplay between the two dominant modes of the nuclear interaction. In the later reaction processes, partial fusion of the projectile with the target nucleus takes place, leading to the formation of a "hot" metastable incompletely fused composite system

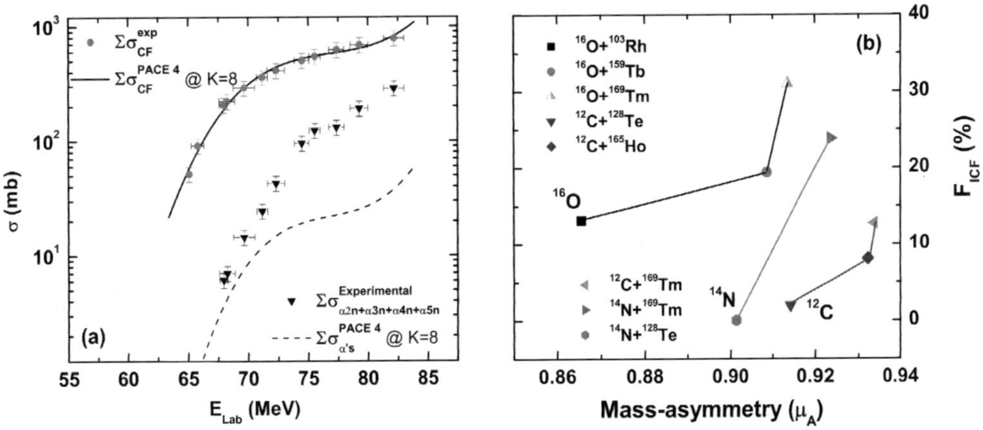


**Figure 1.** (Color Online) (a) Experimentally measured total CF cross section (filled bullets), total $\alpha$'s cross section (filled triangles). A comparison with PACE4 (k = 8) channel has also been shown. (b) The values of $F_{ICF}$ for different projectile-target combinations as a function of entrance channel mass-asymmetry ($\mu_A$) at a constant relative velocity (i.e., $v_{rel} \approx 0.053c$). The lines drawn through the data points guide the eyes for individual ($^{12}C$, $^{14}N$, $^{16}O$) projectiles.

with less mass, charge, and excitation energy as compared to the CF population, where the entire projectile merges with the target nucleus [6, 7, 8, 9]. Though several models and theories [10, 11, 12, 13] are available to understand the dynamics of ICF processes, but no satisfactory explanations of ICF data could be made at low energies. In some recent reports, a large fraction of ICF has been observed at these energies [14, 15, 16, 17, 18, 19], which triggered the resurgent interest to understand the low energy ICF reaction dynamics.

In the present work, the effect of projectile structure (in terms of the $Q_\alpha$ value and projectile dependent mass-asymmetry), and the charge dependence (i.e., the product of the projectile atomic number ($Z_P$) and the target atomic number ($Z_T$)) on the onset and strength of ICF have been studied via excitation function measurements (EFs). On the other hand, the usefulness of ICF as a spectroscopic tool to populate high spin states in the final reaction products has been studied via spin distribution measurements (SDs). The data were obtained using 15 UD Pelletron at the Inter University Accelerator Center (IUAC), New Delhi, India. Irradiations for the EFs were performed in the General Purpose Scattering Chamber [20]. After the irradiations, the radioactivity produced in each target catcher assemblies was measured offline for several days with two high resolution High Purity Germanium (HPGe) detectors. On the other hand, an online experiment to measure the SDs associated to the reaction residues was performed in the Gamma Detector Array-Charge Particle detector Array [21, 22]. The prompt $\gamma$-rays were recorded with 12 Compton suppressed HPGe detectors mounted at angles $45^0$, $99^0$, $153^0$ with respect to the beam axis, and were in coincidence with the particles & $\alpha$'s detected with 14 Phoswich detectors housed in a 14 cm diameter scattering chamber. Details of the experiments (offline and online) can be found elsewhere [23, 24].

The present paper is organized as follows: Interpretation of fusion excitation functions in context of the statistical model code PACE4 [25] is given in Section 2, while Section 3 deals with the spin distribution measurements (SDs) using a particle-gamma coincidence technique, and to obtain the correlation with the driving angular momentum involved in the ICF channels. Section 4 deals with the summary and conclusions of the present work.

## 2. Excitation Function Measurements: Strength and onset of incomplete fusion

The fraction of ICF ($F_{ICF}$) has been measured for several systems [7, 8, 15, 17, 18, 26] using the novel activation technique followed by offline $\gamma$-ray spectroscopy. It may be mentioned that the $F_{ICF}$ has been deduced from the analysis of experimental excitation functions in the framework of statistical model code PACE4. The detailed discussion on data reduction procedure and obtained results can be found elsewhere [15, 26]. As a representative case, the sum of all experimentally identified ICF-channels (filled triangles) is plotted with respect to bombarding energy along with the sum of all CF-channels (solid circles) in Fig. 1(a) for the $^{14}$N+$^{169}$Tm system. The Fig. 1(a) shows the sum of cross-sections for the xn and pxn reactions (solid circles) extracted experimentally and matches satisfactorily with the predictions of PACE4 using the level density parameter constant 'K'=8 (solid line). It may be relevant to mention that the code PACE4 takes into account the excitation energy dependence of the level density parameter (a=A/K) using the prescription of Kataria, Ramamurthy, and Kapoor [27]. In order to trace the suitable value of 'a', to reproduce the experimental fusion EFs, the values of 'K' were varied from 8 to 12. It has been observed that the value 'K'=8 satisfactorily reproduces experimental fusion EFs and the same is highlighted with solid line (See Fig. 1(a)). The value of 'K'=8 was also suggested by Gilbert and Cameron [28] at the studied energy range in this mass region. Further, the sum of all identified $\alpha$-channels cross-sections (filled triangles) is compared with the theoretical $\alpha$-channels cross-sections (dotted line) calculated using PACE4 at the same value of 'K' and the same is presented in Fig. 1(a). As can be seen from the figure, the experimentally deduced $\alpha's$ are higher when compared to corresponding PACE4 predictions ($\Sigma\sigma_{\alpha's}^{PACE4}$). This enhancement over the theoretical prediction may be attributed due to the contribution from ICF. Further, an attempt has been made to deduce experimental ICF cross section ($\Sigma\sigma_{ICF}^{exp}$) by subtracting the experimentally deduced $\alpha's$ with corresponding PACE4 predictions (dashed line). Moreover, for better insights into the onset and strength of ICF, the percentage fraction of ICF (i.e. $F_{ICF}$) is derived using following expression;

$$F_{ICF} = \frac{\Sigma\sigma_{ICF}^{exp}}{\Sigma\sigma_{TF}} \times 100 \quad (1)$$

where, $\Sigma\sigma_{ICF}^{exp}$ and $\Sigma\sigma_{TF}$ are experimental ICF and total fusion cross-sections, respectively. In order to see the dependence of ICF on entrance channel mass asymmetry $\mu_A$(=$A_T$/($A_T$+$A_P$), where $A_P$ is the atomic mass number of the projectile and $A_T$ is the atomic mass number of the target nucleus), the present results are analyzed within the framework of Morgensterns systematics [29, 30]. According to Morgensterns mass-asymmetry systematics, ICF contributes significantly above relative velocity $v_{rel}$(= 0.06 (6 % of c)) of the projectile, and the fraction of ICF ($F_{ICF}$) should increase with the mass-asymmetry ($\mu_A$) of the system. The $F_{ICF}$ for our recently performed experiment ($^{14}$N+$^{169}$Tm system [26]) is compared with the available nearby systems [see references in [26]] and are plotted in Fig. 1(b) at a constant relative velocity ($v_{rel}$ $\approx$ 0.053c). Lines through the data points are drawn to guide the eyes for individual projectiles. As can be seen from this figure, the variation of $F_{ICF}$ does not increases with the $\mu_A$ of the system, however, the value of $F_{ICF}$ increases with $\mu_A$ for individual projectiles. One of the reasons for such behavior can be understood by considering that, for a given projectile, as the mass asymmetry increases the Coulomb effect (repulsion) also increases, giving rise to a larger breakup probability. In order to display the charge dependence of the ICF strength function, the $F_{ICF}$ values are plotted in Fig. 2(a) as a function of $Z_PZ_T$ of the system along with $F_{ICF}$ of other systems available in the literature [[26], references therein] at the same $v_{rel}$ (i.e. $\approx$ 0.053c). As can be seen from the figure, the $F_{ICF}$ follows almost a linear growth as the charge product

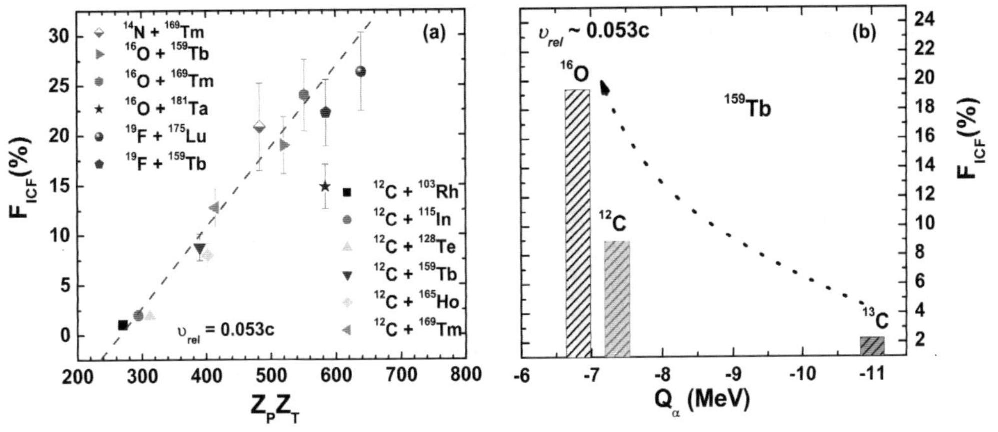


**Figure 2.** (Color Online) (a) Incomplete fusion strength function ($F_{ICF}$) of various systems as a function of $Z_PZ_T$. The dashed line is drawn to guide the eyes. (b) Comparison of $F_{ICF}$ on the basis of $Q_\alpha$ at a constant $v_{rel} \approx 0.053c$.

$Z_PZ_T$ increases, which indicates the role of Coulomb repulsion on ICF fraction. Hence, it may be concluded that an increase in the value of $Z_PZ_T$ of the system enhances the strength of the Coulomb interaction resulting in the larger breakup probability. Moreover, after incorporating the Coulomb barrier of a given projectile-target combination in calculations, it has been observed that ICF contributes significantly even at slightly above barrier energies, and the values of $F_{ICF}$ are different for different projectile-target combinations, which reveals a projectile and/or target dependence on ICF fraction (for details see ref. [26]).

For better understanding of underlying approaches for the individual projectiles, the percentage fraction of incomplete fusion ($F_{ICF}$) is analyzed in terms of projectiles $Q_\alpha$ [31]. The values of $F_{ICF}$ for the $^{12}$C,$^{13}$C,$^{16}$O+$^{159}$Tb are plotted as a function of $Q_\alpha$ in Fig. 2(b) at a constant $v_{rel}$ = 0.053c. As shown in this figure, the probability of ICF for the $^{13}$C projectile is noticeably smaller than that for the $^{12}$C projectile. The strikingly different ICF fractions for $^{13}$C and $^{12}$C induced reactions point towards the strong projectile structure effect. It may be pointed out that $^{12}$C is a well known $\alpha$-cluster nucleus with $Q_\alpha$ = -7.37 MeV. On the other hand, $^{13}$C has a relatively larger negative $Q_\alpha$ (= -10.64 MeV) than $^{12}$C. The large negative $Q_\alpha$ for $^{13}$C translates into the smaller breakup probability into constituent $\alpha$-clusters, resulting in a smaller ICF fraction than that found in $^{12}$C induced reactions. It is evident in Fig.2b that the probability of ICF is found to be smaller for larger negative $Q_\alpha$ projectiles. For example, the value of $F_{ICF}$ for the $^{16}$O ($Q_\alpha$ =7.16 MeV)+ $^{159}$Tb system is found to be 19% which is reduced to only 3% for the $^{13}$C ($Q_\alpha$ =-10.64 MeV)+ $^{159}$Tb system. The same systematics follow for other targets and can be found elsewhere [15, 32]. Hence, it can be inferred that the $Q_\alpha$ can be taken as an important entrance channel parameter which essentially dictates the probability of ICF.

## 3. Spin Distribution Measurements: localization of the $\ell$ window

In head-on or near head-on heavy ion collisions, the reacting nuclei may lead to the formation of a compound nucleus (CN), where the driving input angular momentum '$\ell$' associated can reach

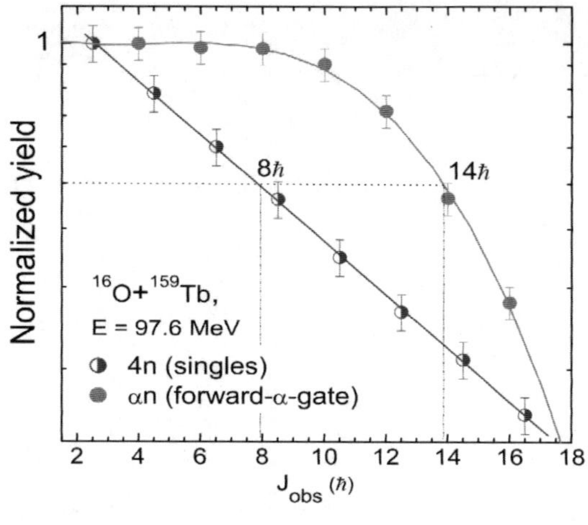


**Figure 3.** (Color Online) Experimentally generated spin-distributions of $\alpha$n and 4n-channels. Lines and curves are the best fit to the experimental data points [40]. Fitting and normalization procedures are explained in the [41].

upto limiting angular momentum value or critical angular momentum of the system '$\ell_{crit}$' [33]. Consequently, the target nucleus hugs the projectile with the involvement of all nucleonic degree of freedoms leading to the formation of a fully equilibrated CN with high excitation energy ($E^*$). Thus, CF residues are formed at high $E^*$ and low '$\ell$' (i.e. $\ell \leq \ell_{crit}$). On the other hand, the probability of CF is expected to be negligible for '$\ell$' greater than '$\ell_{crit}$' (as per the sharp cutoff approximation) [34, 35, 36, 37], and thus CF gives way to ICF, where residues achieve low $E^*$ (due to the involvement of partial degrees of fusion) and high $\ell$ (due to relatively higher impact parameter). In such a case, the number of 'yrast'-like transitions are much larger than that of statistical transitions. As a result, the spin-distributions of CF and ICF residues are expected to be entirely different in nature and can be used as a sensitive tool to probe the mode of reaction by looking into the spin population of the entry state. In order to probe the role of high $\ell$-values in the production of ICF residues, the spin-distributions of various CF and ICF channels have been measured [38, 24, 39, 40, 41]. The details of the online experiment can be found elsewhere [38].

As a representative case, the spin-distributions for the 4n and $\alpha$n channels, that is, $^{171}$Ta(4n) identified from singles spectra, and $^{170}$Lu($\alpha$n) identified from fast-$\alpha$-gated-$\gamma$ spectra detected in the forward cone are plotted in Fig. 3. As can be seen from this figure, the trend of the yield distribution for the $\alpha$-emitting channel identified from forward $\alpha$-gated-$\gamma$ spectra is distinctly different from that of the 4n-channel identified from singles data. A gradual steep rise in yield for 4n-channel towards the band head indicates continuous spin population for a broad spin range during the de-excitation. However, in case of $\alpha$n-channel, the intensity increases up to a value of observed spin ($J_{obs}$) $\approx$ 10 $\hbar$ and then shows a constant behavior towards the band head. Thus, the constant behavior from 8 $\hbar$ to 2$\hbar$, reveals the hindrance in the population of low lying states in ICF processes. Moreover, the spin-distribution patterns of neutron and $\alpha$-emitting channels clearly suggest reaction dependent de-excitation patterns. In order to reinstate distinct nature of de-excitation pattern observed in the Fig. 3, the spin distributions for the $^{168}$Lu residue identified from forward $\alpha$-gated-$\gamma$ spectra (F) and backward $\alpha$-gated-$\gamma$ spectra (B) are shown in Fig. 4 (a and b) for the projectile energies, $E_{Lab}$ = 83.5 $\pm$ 1.5, 88.5 $\pm$ 1.5, 93.5 $\pm$ 1.5 and 97.6 $\pm$ 1.4 MeV, respectively. As can be observed from Fig. 4, the trend of spin-distributions

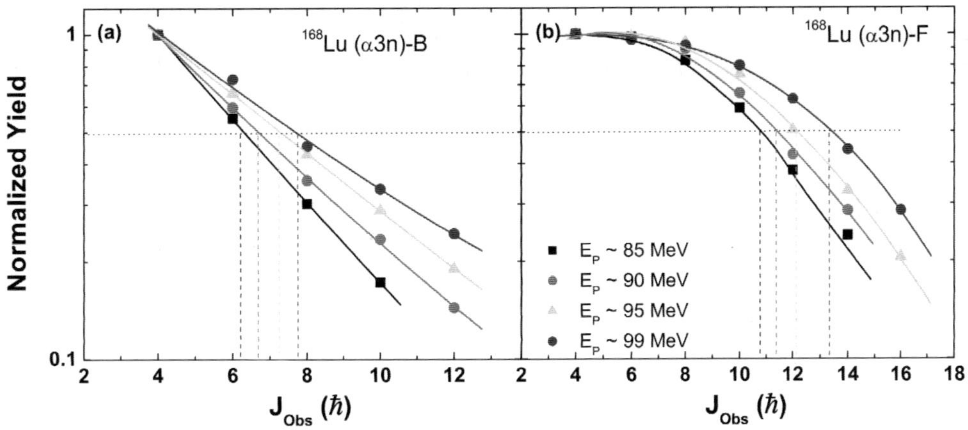


**Figure 4.** (Color Online) Experimentally generated spin-distributions of (a) $\alpha$3n-B, and (b) $\alpha$3n-F channels. Lines and curves are the best fit to the experimental data points [38].

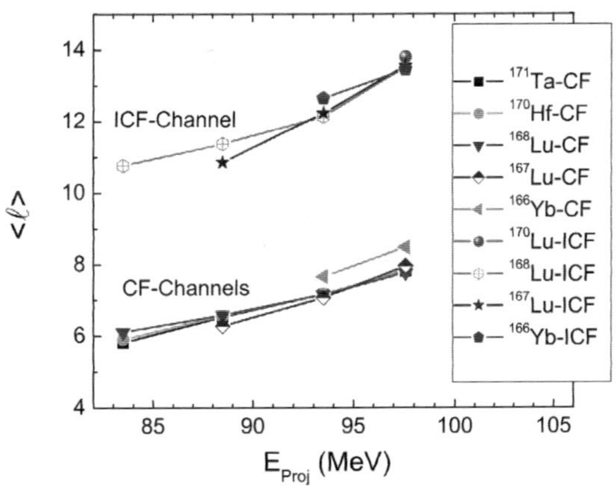


**Figure 5.** (Color Online) The value of mean input angular momenta $< \ell >$ associated with the production of different isotopes in $^{16}$O+$^{159}$Tb system at studied energies. Lines and curves through data points are drawn to guide the eyes.

for $^{168}$Lu-B(complete fusion) and $^{168}$Lu-F (incomplete fusion) is visibly different, and follow the same behavior as that of 4n- and $\alpha$n-channels, respectively. It may be mentioned the for $\alpha$xn-F-channels (ICF residues), the intensity appears to be almost constant up to a certain value of $J_{obs}$, and varies with the incident projectile energies, which then decreases towards band head. From the experimental observations it may be concluded in case of ICF, population of low lying states or lower levels are strongly hindered and there is abscence of feeding of $\gamma$'s from higher levels.

In order to understand the usefulness of ICF as a tool to populate high spin states in the final reaction products, the values of $\ell$ involved in CF- and ICF-channels have been deduced from the analysis of experimental spin-distributions [38, 24] and are plotted in Fig. 5 as a function of incident projectile energy. As indicated in this figure, the $< \ell >$ values involved in complete fusion (xn/$\alpha$xn) and incomplete fusion ($\alpha$xn) channels are found to be $\approx 6\ \hbar$ and $\approx 11$ $\hbar$respectively, at projectile energy $\approx 83.5 \pm 1.5$ MeV, and increases linearly for higher energies. On comparing the $< \ell >$ values involved in the ICF ($\approx 11\ \hbar$) at the lowest projectile energy is found to be 38% than that involved in the CF channels ($\approx 8\ \hbar$) at the highest experiential incident laboratory energy. This clearly demonstrates the involvement of higher $\ell$-values in the production of incomplete channels which may lead to the population of higher spin-states in the final reaction products which may not be possible otherwise.

## 4. Summary and Conclusions

The present paper gives the experimental results for the excitation function and particle-$\gamma$ coincidence experiments performed using 15UD pelletron facility at the IUAC, New Delhi, India. In case of excitation function experiments, the cross sections of $\alpha$ exit channels were found to be enhanced when compared with the theoretical model using code PACE4. A comparison of $F_{ICF}$ for different projectile and on the same target ($^{169}$Tm) combinations displays higher ICF probability for the $^{16}$O+$^{169}$Tm system, though it is a less mass-asymmetric system than $^{12}$C, $^{14}$N+$^{169}$Tm systems (see Fig. 1(b)). Morgensterns mass-asymmetry systematics do not explain the experimental data of eight nearby systems as a whole, however, the value of $F_{ICF}$ is found to increase with entrance channel mass-asymmetry $\mu_A$ for individual projectiles. Further, a strong dependence of the incomplete fusion strength function ($F_{ICF}$) on the Coulomb effect, i.e., $Z_PZ_T$ was observed on comparing systems available in literature. Moreover, the effect of $Q_\alpha$ on ICF fraction has been observed for strongly bound projectiles. The $F_{ICF}$ has been found to decrease for large negative $Q_\alpha$ of the projectiles. If confirmed for other projectile-target combinations, this may provide an important input to understand the complex ICF dynamics at low incident energies.

In the case of spin distributions, residues populated through the ICF routes are found to be originated from the narrow spin population, localized near and/or above to the critical angular momentum for CF to occur, where a given direct-$\alpha$-fragment is emitted to release excess driving angular momenta. This reveals a competition from successively opened ICF channels for each value of $\ell$ above the $\ell_{crit}$ for normal fusion (CF). Further, results based on spins reveals that the high $\ell$-values associated with peripheral interactions contribute to open up direct-$\alpha$-channels.

In conclusion, to achieve better understanding of the dependence of low-energy incomplete fusion reaction dynamics on various entrance channel parameters, more precise and diversified data is required.

## 5. Acknowledgments

The authors thank the Director of the Inter University Accelerator Center (IUAC), New Delhi, India for extending experimental facilities. One of the authors (V.R.S.) thanks CONACYT for providing financial support under Project No. CB-01-254619.

## References

[1] Aguilera E F *et al* 2011 Phys. Rev. Lett. **107** 092701
[2] Aguilera E F *et al* 2016 Phys. Rev. C **93** 034613
[3] Guimaraes V *et al* 2016 Phys. Rev. C **93** 064607
[4] Bruno M *et al* 2013 Eur. Phys. Jour. A **49:** 128
[5] Fabris D *et al* 1994 Phys. Rev. C (R) **50(3)** 1261
[6] Sharma V R *et al* 2011 Phys. Rev. C **84** 014612
[7] Yadav A *et al* 2012 Phys. Rev. C **85** 064617

[8] Shuaib M *et al* 2016 Phys. Rev. C **94** 014613
[9] Singh P P *et al* 2015 Journal of Physics: Conference Series **590** 012031
[10] Hussein M S, Pato M P, Canto L F, and Donangelo R 1992 Phys. Rev. C **46** 377 ibid. 1995 Nucl. Phys. A **85c** 588; Hussein M S and Toledo Piza A F R 1994 Phys. Rev. Lett. **72** 2693
[11] Yabana K 1997 Prog. Theor. Phys. **97** 437
[12] Diaz-Torres A and Thompson I J 2002 Phys. Rev. C **65** 024606
[13] Kolata J J *et al* 1998 Phys. Rev. Lett. **81** 4580
[14] Sharma M K *et al* 2015 Phys. Rev. C **91** 024608
[15] Sharma V R *et al* 2014 Phys. Rev. C **89** 024608
[16] Singh P P *et al*. 2007 Eur. Phys. J. A **34** 2939
[17] Yadav A *et al* 2017 Phys. Rev. C **96** 044614
[18] Shuaib M *et al* 2017 J. Phys. G: Nucl. Part. Phys. **44** 105108
[19] Singh D P *et al* 2015 EPJ Web of Conferences **86** 00050
[20] Puttaswamy N G, Badiger N M, and Raja Rao M 1991 DAE symposium on nuclear physics Contributed papers. Vol. **34B** 405
[21] Muralithar S *et al* 2013 Nucl. Instrum. Methods Phys. Res. A **729** 849855
[22] Muralithar S *et al* 2014 Pramana **82** 769
[23] Singh P P *et al* 2008 Phys. Rev. C **77** 014607
[24] Sharma V R *et al* 2015 J.Phys.G,Nucl.Part.Phys.**42** 055113
[25] Gavron A 1980 Phys. Rev. C **21** 230
[26] Kumar R *et al* 2017 Phys. Rev. C **96** 054614
[27] Kataria S K, Ramamoorthy V S and Kapoor S S 1978 Phys. Rev. C **18** 549
[28] Gilbert A and Cameron A G N 1965 Can. J. Phys. **43** 1446
[29] Morgenstern H, Bohne W, Galster W, Grabisch K and Kyanowski A 1984 Phys. Rev. Lett. **52** 1104
[30] Morgenstern H *et al* 1982 Phys. Lett. B **113** 463
[31] Yadav A *et al* 2012 Phys. Rev. C **86** 014603
[32] Babu K S *et al* 2003 J. Phys. G **29** 1011
[33] Plasil F 1980 Journal de Physique Colloques **C10** 41
[34] Barker J H *et al* 1980 Phys. Rev. Lett. **45** 424
[35] Parker D J *et al* 1989 Phys. Rev. C **39** 2256
[36] Wu J R, Lee I Y 1980 Phys. Rev. Lett. **45** 8
[37] Hojman D *et al* 2006 Phys. Rev. C **74** 044604
[38] Sharma V R *et al* 2016 Nuclear Phys. A **946** 182-193
[39] Sharma V R *et al* 2017 EPJ Web of Conferences **163** 00054
[40] Singh P P *et al* 2009 Phys. Rev. C **80** 064603
[41] Singh P P *et al* 2009 Phys. Lett. B **671** 20

IOP Conf. Series: Journal of Physics: Conf. Series **1078** (2018) 012021 doi:10.1088/1742-6596/1078/1/012021

# Structure of baryon states from non-perturbative methods.

**T. Yépez-Martínez[1], O. Civitarese[1], D. A. Amor Quiroz[2] and P. O. Hess[2]**

[1] Departamento de Física, Universidad Nacional de La Plata, C.C.67 (1900), La Plata, Argentina

[2] Instituto de Ciencias Nucleares, UNAM, Circuito Exterior, C.U., A.P. 70-543, 04510 México, D.F., México.

E-mail: yepez@fisica.unlp.edu.ar, osvaldo.civitarese@fisica.unlp.edu.ar, arturo.amor@nucleares.unam.mx, hess@nucleares.unam.mx

**Abstract.** In this approach we have adapted non-perturbative many-body methods to describe baryon-like states built from effective quarks, antiquarks degrees of freedom and collective RPA states. As starting point, we have used the QCD Hamiltonian written in the Coulomb gauge, and considered a confining interaction to describe the non-perturbative effects of dynamical gluons. The effective Hamiltonian used is expressed in terms of the effective quark-antiquark, di-quarks and di-antiquark excitations and approximately diagonalized by mapping quark-antiquark pairs and di-quarks (di- antiquarks) onto phonon states. For the structure of the vacuum of the theory, color-scalar- and color-vector-structures are introduced to account for ground state correlations. We present the structure of the baryon-like states, beyond the common three constituent quarks, describe their interactions and present a simple calculation to illustrate the method.

## 1. Introduction

The low-energy QCD regime has been widely studied during the last four decades. Several aspects have been investigated by the implementation of non-perturbative methods, among them the two more outstanding are the Lattice gauge theory (LGT) and the Dyson-Schwinger Equations (DSEs). The LGT has had several and important advances in the understanding of fundamental phenomena dominated by the low-energy regime of the theory, the two more notorious are: a first principle description of the confinement potential of quarks and gluons [1] and the chiral symmetry breaking. The LGT has provided a satisfactorily description of the low energy meson and baryon spectrum [2–4], but due to the computational limitations, the pion mass used is considerable larger than the experimental one [3] and the accessibility to higher excited states is complicated. The DSEs has had also important and significant advances in the understanding of the low-energy regime of QCD. The description of the dynamical quark and gluon mass generation by the implementation of self-consistent techniques has been outstanding, as well as its contributions to the description of the hadron spectrum [5]. Particularly, the DSEs results for the low energy baryon spectrum is able to describe properly certain aspects of the spectrum, *e.g.*, the Roper resonance description as a radial excitation of the nucleon and its low mass understanding as an effect of a meson cloud that shields the quark core [6]. The models based on the identification of effective degrees of freedom and their interactions has been also a crucial step in order to describe physical systems like mesons and baryons. In this sense, the many-body methods [7–9] have been a guide during a long time, achieving remarkable success in nuclear and hadron physics. Despite all of these efforts, the low-energy regime of QCD remains unsolved and still more work has to be done in order to get more insights about the physics of different phenomena involved in the proper description of physical states and the interactions between quarks and gluons inside them.

Here, we shall show that the identification of effective degrees of freedom and the use of the *Random-Phase-Approximation* (RPA) many-body method, implemented in the non-perturbative regime of QCD, can generate a feasible description of the interactions between effective quarks and antiquarks with collective RPA particle-hole ($ph$), di-quark (di-$q$) and di-antiquark ($di - \bar{q}$) systems. With all of them a description of the baryon states beyond the common three quarks can be provided. The results look promising and encouraging for further investigation along this line. In this work, the identification of single-particle and hole excitations as well as collective degrees of freedom allows for the construction of effective interactions and discussed their structure in terms of the various couplings allowed by the QCD Hamiltonian in the Coulomb gauge [10]. Several features of the non-perturbative regime of QCD have been explored within the Coulomb gauge framework [11–17]. In [21], the sensitivity of the chiral symmetry breaking has been explored in the excited baryon spectrum, within the Coulomb gauge QCD formalism. Such analysis has been performed in a controlled way in terms of the number of harmonic oscillator (h.o.) shells, similar to that performed in [9, 22].

The present description is based on the use of effective quark and antiquark states and quark-antiquark, di-$q$ and di-$\bar{q}$ pairs and their interactions. We start from the QCD Hamiltonian written in the Coulomb gauge [12, 13] and perform a diagonalization of the kinetic energy term, expressed in the h.o. basis, to get quark and antiquark states. With the effective single-quark and antiquark states, we rewrite the color-charge density interactions and replace the instantaneous color Coulomb interaction by a static potential $V(r) = (-\alpha/r) + (\beta r)$ [9]. A further diagonalization of the pair-like terms ($ph$, di-$q$ and di-$\bar{q}$) of the Hamiltonian is performed by applying the RPA method [23], and finally, the remaining terms of the Hamiltonian are treated as a generalized phonon-single-particle- and pair-interactions, keeping in all cases the symmetries of the QCD degrees of freedom at the level of meson and baryon-like states. The main characteristic of the method consist in the use of coloured-pairs configurations of both quarks and antiquarks, in addition to the colorless quark-antiquark pair configurations, to construct baryon-like states by coupling these configurations to quarks or antiquarks. The formalism and general aspects of the model are described in Sections 2, a detailed description of the interactions is presented in Section 4 and the application of the method is illustrated in Section 4.1. The conclusions are drawn in Section 5.

## 2. General aspects and the effective Hamiltonian.

### *2.1. QCD Hamiltonian in the Coulomb gauge.*

As we have mentioned in the Introduction, we start from the QCD Hamiltonian in its canonical Coulomb gauge representation [10],

$$\begin{aligned} \boldsymbol{H}^{QCD} &= \int \left\{ \frac{1}{2} \left[ \mathcal{J}^{-1} \boldsymbol{\Pi}^{tr} \cdot \mathcal{J} \boldsymbol{\Pi}^{tr} + \mathcal{B} \cdot \mathcal{B} \right] - \overline{\psi} \left( -i\gamma \cdot \nabla + m \right) \psi - g \overline{\psi} \gamma \cdot A \psi \right\} d\mathbf{x} \\ &+ \frac{g^2}{2} \int \mathcal{J}^{-1} \rho^c(\mathbf{x}) \langle c, \mathbf{x} | \frac{1}{\nabla \cdot \mathcal{D}} (-\nabla^2) \frac{1}{\nabla \cdot \mathcal{D}} | c' \mathbf{y} \rangle \mathcal{J} \rho^{c'}(\mathbf{y}) d\mathbf{x} d\mathbf{y}, \end{aligned} \tag{1}$$

which has been widely studied in the past in its non-perturbative regime [7,9,12–17,21,22,24,25], in order to get insights about the confinement of color-charged particles along with the constituent quark and gluon masses. Also several bound-states like meson, baryons and others predicted by the theory, *e.g.* hybrids and glueballs, has been described [7, 8, 18–22, 26, 27].

The Hamiltonian of Eq. (1) takes into account the interactions between quarks and gluons through the QCD *Instantaneous color-Coulomb Interaction* (QCD-IcCI) between color-charge-densities of quarks and gluons. At low energy the effects of dynamical gluons in the QCD-IcCI are accounted for by the interaction $V(|\mathbf{x}-\mathbf{y}|) = -\frac{\alpha}{|\mathbf{x}-\mathbf{y}|} + \beta|\mathbf{x}-\mathbf{y}|$, which is obtained from a self-consistent treatment of the last term of the Hamiltonian [12, 13].

The quark sector of the Hamiltonian of Eq. (1), is given by

$$\begin{aligned} \boldsymbol{H}^{QCD}_{eff} &= \int \left\{ \boldsymbol{\psi}^\dagger(\mathbf{x})(-i\boldsymbol{\alpha}\cdot\boldsymbol{\nabla} + \beta m)\boldsymbol{\psi}(\mathbf{x}) \right\} d\mathbf{x} - \frac{1}{2}\int \rho_c(\mathbf{x}) V(|\mathbf{x}-\mathbf{y}|)\rho^c(\mathbf{y}) d\mathbf{x}d\mathbf{y} \\ &= \boldsymbol{K} + \boldsymbol{H}_{\text{Coul}} , \end{aligned} \quad (2)$$

where $\rho^c(\mathbf{x}) = \psi^\dagger(\mathbf{x})T^c\psi(\mathbf{x})$ is the quark and antiquark color-charge density. In Eq. (2), the first term is the kinetic energy, while the second term is the QCD-IcCI in its simplified form. The fermion field $\psi^\dagger(\mathbf{x})$, whose quantization was explained in [9], is expanded in terms of creation and annihilation operators of particles (antiparticles) in a basis of h.o. functions.

The use of the h.o. basis has several advantages, like a considerable simplification in the matrix elements and their calculation, also the use of the h.o. basis represents finite range fields which is something expected for quarks and antiquarks. However, its use requires a pre-diagonalization of the kinetic term, which is performed by means of an unitary transformation between creation (annihilation) operators in the h.o. basis $\boldsymbol{q}^\dagger_{\tau(Nl)}(\boldsymbol{q}^{\tau(Nl)})$ and those belonging to an effective basis $\boldsymbol{Q}^\dagger_{\lambda k\pi_q}(\boldsymbol{Q}^{\lambda k\pi_q})$ of quarks, such that

$$\boldsymbol{q}^\dagger_{\tau(Nl)J_qC_q(Y_q,T_q),M_{J_q}M_{C_q}M_{T_q}} = \sum_{\lambda\pi_q k}\left(\alpha^{J_q,T_q}_{\tau(Nl),\lambda\pi_q k}\right)^* \boldsymbol{Q}^\dagger_{\lambda k\pi_q J_qC_q(Y_q,T_q),M_{J_q}M_{C_q}M_{T_q}}\,\delta_{\pi_q,(-1)^{\frac{1}{2}-\tau+l}} . \quad (3)$$

The effective quark $\boldsymbol{b}^\dagger$ and antiquark $\boldsymbol{d}$ operators are obtained via

$$\boldsymbol{Q}^\dagger_{\frac{1}{2}k\pi_q J_qC_q(Y_q,T_q),M_{J_q}M_{C_q}M_{T_q}} = \boldsymbol{b}^\dagger_{k\pi_q J_qC_q(Y_q,T_q),M_{J_q}M_{C_q}M_{T_q}} . \quad (4)$$

and

$$\boldsymbol{Q}^\dagger_{-\frac{1}{2}k\pi_q J_qC_q(Y_q,T_q),M_{J_q}M_{C_q}M_{T_q}} = \boldsymbol{d}_{k\pi_q J_qC_q(Y_q,T_q),M_{J_q}M_{C_q}M_{T_q}} . \quad (5)$$

The sub-index $\pi_q = \pm$ indicates the parity of the effective quark or antiquark and $k = 1, 2, \cdots$ runs over all orbital states. The index $J_q = \frac{1}{2}, \frac{3}{2}, \cdots$ indicates the total $(J_q = l \pm \frac{1}{2})$ single-particle spin. The flavor hypercharge and isospin quantum numbers for quarks are given by $(Y_q, T_q) = \left(\frac{1}{3}, \frac{1}{2}\right), \left(-\frac{2}{3}, 0\right)$. The quarks and antiquarks belong to a triplet $C_q = (10)$ and anti-triplet $\bar{C}_q = C_{\bar{q}} = (01)$ color irreducible representations (irreps), respectively, which are conjugate representations. From now on, we will use the short-hand notation $\gamma_q = \{J_q, C_q, (Y_q, T_q)\}$ for the quarks and $\bar{\gamma}_q = \{J_q, \bar{C}_q, (\bar{Y}_q, T_q)\}$ for antiquarks irreps with $\bar{Y}_q = -Y_q$, while $\mu_q = \{M_{J_q}, M_{C_q}, M_{T_q}\}$ and $\bar{\mu}_q = \{\bar{M}_{J_q}, \bar{M}_{C_q}, \bar{M}_{T_q}\}$ will be used for the magnetic numbers with $\bar{M}_{J_q} = -M_{J_q}$ and $\bar{M}_{T_q} = -M_{T_q}$.

The kinetic ($\boldsymbol{K}$) term rewritten in terms of effective quarks and antiquarks operators has the following structure [9]

$$\boldsymbol{K} = \sum_{k\pi_q\gamma_q} \varepsilon_{k\pi_q\gamma_q} \sum_{\mu_q}\left(\boldsymbol{b}^\dagger_{k\pi_q,\gamma_q\mu_q}\boldsymbol{b}^{k\pi_q,\gamma_q\mu_q} - \boldsymbol{d}^{k\pi_q,\bar{\gamma}_q\bar{\mu}_q}\boldsymbol{d}^\dagger_{k\pi_q,\bar{\gamma}_q\bar{\mu}_q}\right), \quad (6)$$

where the rules to raise and lower indices are taken from [28].

The QCD-IcCI term, in its simplified form ($\boldsymbol{H}_{\rm Coul}$), rewritten in terms of effective quark and antiquark operators is given by [9]

$$\begin{aligned}\boldsymbol{H}_{\rm Coul} &= -\frac{1}{2}\sum_{L}\sum_{\lambda_i \mathbf{q}_i} V^{L}_{\{\lambda_i \mathbf{q}_i\}}\Big( \left[\mathcal{F}_{\lambda_1\mathbf{q}_1,\lambda_2\mathbf{q}_2;\gamma_{f0}}\mathcal{F}_{\lambda_3\mathbf{q}_3,\lambda_4\mathbf{q}_4;\bar{\gamma}_{f0}}\right]^{\gamma_0}_{\mu_0} + \left[\mathcal{F}_{\lambda_1\mathbf{q}_1,\lambda_2\mathbf{q}_2;\gamma_{f0}}\mathcal{G}_{\lambda_3\mathbf{q}_3,\lambda_4\mathbf{q}_4;\bar{\gamma}_{f0}}\right]^{\gamma_0}_{\mu_0} \\ &+ \left[\mathcal{G}_{\lambda_1\mathbf{q}_1,\lambda_2\mathbf{q}_2;\gamma_{f0}}\mathcal{F}_{\lambda_3\mathbf{q}_3,\lambda_4\mathbf{q}_4;\bar{\gamma}_{f0}}\right]^{\gamma_0}_{\mu_0} + \left[\mathcal{G}_{\lambda_1\mathbf{q}_1,\lambda_2\mathbf{q}_2;\gamma_{f0}}\mathcal{G}_{\lambda_3\mathbf{q}_3,\lambda_4\mathbf{q}_4;\bar{\gamma}_{f0}}\right]^{\gamma_0}_{\mu_0}\Big), \qquad (7)\end{aligned}$$

where we have compacted the single particle orbital number, parity and irreps, into the short-hand notation $\mathbf{q}_i = k_i \pi_{q_i} \gamma_{q_i}$, and use for the (flavorless) quantum numbers of the intermediate coupling in the interaction the label $\gamma_{f_0} = \{L, (11), (0,0)\}$ and for their magnetic projections $\mu_{f_0} = \{M_L, M_C, 0\}$, respectively. The conjugate representations satisfy $\bar{\gamma}_{f_0} = \gamma_{f_0}$ and $\bar{\mu}_{f_0} = \{-M_L, \bar{M}_C, 0\}$. Then for the total couplings (upper index) and magnetic numbers (lower index) of the interaction, we have used $\gamma_0 = \{0, (00), (0,0)\}$ and $\mu_0 = \{0,0,0\}$ respectively. The operators $\mathcal{F}$ and $\mathcal{G}$ are written

$$\begin{aligned}\mathcal{F}_{\lambda_1\mathbf{q}_1,\lambda_2\mathbf{q}_2;\gamma_{f0},\mu_{f0}} &= \frac{1}{\sqrt{2}}\left\{\delta_{\lambda_1,\frac{1}{2}}\delta_{\lambda_2,\frac{1}{2}}\left[\boldsymbol{b}^{\dagger}_{\mathbf{q}_1}\otimes \boldsymbol{b}_{\bar{\mathbf{q}}_2}\right]^{\gamma_{f0}}_{\mu_{f0}} - \delta_{\lambda_1,-\frac{1}{2}}\delta_{\lambda_2,-\frac{1}{2}}\left[\boldsymbol{d}_{\bar{\mathbf{q}}_1}\otimes \boldsymbol{d}^{\dagger}_{\bar{\mathbf{q}}_2}\right]^{\gamma_{f0}}_{\mu_{f0}}\right\} \\ \mathcal{G}_{\lambda_1\mathbf{q}_1,\lambda_2\mathbf{q}_2;\gamma_{f0},\mu_{f0}} &= \frac{1}{\sqrt{2}}\left\{\delta_{\lambda_1,-\frac{1}{2}}\delta_{\lambda_2,\frac{1}{2}}\left[\boldsymbol{d}_{\mathbf{q}_1}\otimes \boldsymbol{b}_{\bar{\mathbf{q}}_2}\right]^{\gamma_{f0}}_{\mu_{f0}} - \delta_{\lambda_1,\frac{1}{2}}\delta_{\lambda_2,-\frac{1}{2}}\left[\boldsymbol{b}^{\dagger}_{\mathbf{q}_1}\otimes \boldsymbol{d}^{\dagger}_{\bar{\mathbf{q}}_2}\right]^{\gamma_{f0}}_{\mu_{f0}}\right\}. \qquad (8)\end{aligned}$$

The $\mathcal{F}^2$ and $\mathcal{G}^2$ terms describe the interactions with two creation and two annihilation operators $H_{22}$, as well as four creation $H_{40}$ and four annihilation operators $H_{04}$, which are the structures relevant for the RPA bosonization. The RPA method leaves out the Hamiltonian terms $H_{31}$ and $H_{13}$, *i.e.* the $\mathcal{F}\mathcal{G}$ and $\mathcal{G}\mathcal{F}$ terms, corresponding to three creation and one annihilation operators and their Hermitian conjugate terms. These terms are taken into account in the mapped Hamiltonian of Section 4.

The Hamiltonian expressed in the effective quark and antiquark operators, Eqs. (6) and (7), describes the interactions of these effective particles in bound states like mesons and baryons. In [9] the low energy meson states with $J^{\pi} = 0^{\pm}, 1^{-}$ were represented by particle(quarks)-hole(antiquarks) (or ph) phonons by means of the *Tamm-Dancoff-Approximation* (TDA) and the RPA methods. The *ph* structure of the interactions in the Hamiltonian are shown in Figure 1a). The TDA and RPA methods provided some insights about the low-energy meson spectrum and the influence of *ph*-correlations in the vacuum. The *ph*-correlations considered in the vacuum

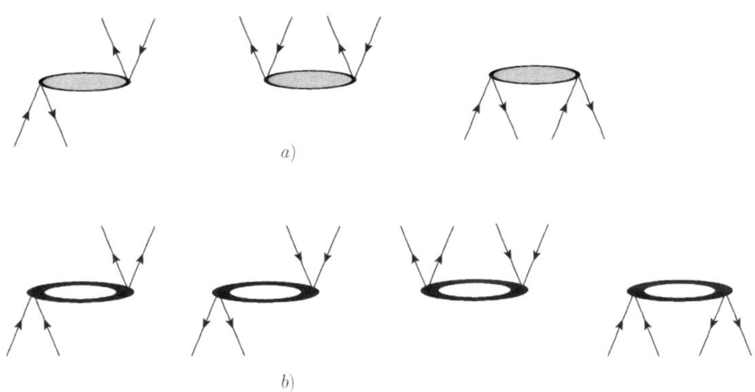


**Figure 1.** $H_{22}$, $H_{40}$ and $H_{04}$ interaction terms for a) particle-hole and b) di-quark and di-antiquarks subspaces, respectively.

were colorless in order to describe meson-states as one-phonon excitations. In the present work, we go further and consider color-correlations in a more general description of the vacuum. These color correlations, viewed as excitations of the vacuum, can not describe physical states but they may be coupled to color-vector configurations to describe bound states *e.g.*, baryon states, which are constructed such that the overall color is zero.

## 3. Basic operators and representations.

In the following, for the vacuum, meson-like (or *ph*-pair), di-$q$ and di-$\bar{q}$ states we use a similar short-hand notation $\gamma_n = \{J_n, C_n, (Y_n, T_n)\}$. The sub-index $n = 0, q, p, a, r$ reads for the vacuum, quarks, mesons, di-$q$ and di-$\bar{q}$ states, respectively. In this work we only consider quarks and antiquarks with total spin $J_q = \frac{1}{2}$. Higher single particle spin states are left for a further publication. For the baryon state representations and magnetic components we will use $\gamma_{B_i} = \{J_{B_i}, C_{B_i}, (Y_{B_i} T_{B_i})\}$ and $\mu_{B_i} = \{M_{J_{B_i}}, M_{C_{B_i}}, M_{T_{B_i})}\}$, respectively.

### *3.1. Vacuum state.*

In [9] the correlated vacuum included colorless *ph* correlations, constructed explicitly by using the RPA formalism, thus the RPA solutions aimed to describe meson-like states. In general, the ground state can include also colored correlations by means of (11)-*ph*-pairs and (20), (01)-di-$q$ and (02), (10)-di-$\bar{q}$ operators. Up to second order in the quark-antiquarks operators, we can relate the new ground state to the vacuum state $|\tilde{0}\rangle$ via

$$
\begin{aligned}
|RPA\rangle &= |\tilde{0}\rangle + \sum_{p\pi_p;\gamma_p} \sum_{\mathbf{q}'_i \mathbf{q}_i} Z^{p\pi_p,\gamma_p}_{(1)\ \mathbf{q}'_1\bar{\mathbf{q}}'_2,\mathbf{q}_1\bar{\mathbf{q}}_2} \times \left[ [\boldsymbol{b}^\dagger_{\mathbf{q}'_1} \otimes \boldsymbol{d}^\dagger_{\bar{\mathbf{q}}'_2}]^{\gamma_p} \otimes [\boldsymbol{b}^\dagger_{\mathbf{q}_1} \otimes \boldsymbol{d}^\dagger_{\bar{\mathbf{q}}_2}]^{\bar{\gamma}_p} \right]^{\gamma_0}_{\mu_0} |\tilde{0}\rangle \\
&+ \sum_{a\pi_a;\gamma_a} \sum_{\mathbf{q}'_i \mathbf{q}_i} Z^{a\pi_a,\gamma_a}_{(2)\ \mathbf{q}'_1\mathbf{q}'_2,\bar{\mathbf{q}}_1\bar{\mathbf{q}}_2} \times \left[ [\boldsymbol{b}^\dagger_{\mathbf{q}'_1} \otimes \boldsymbol{b}^\dagger_{\mathbf{q}'_2}]^{\gamma_a} \otimes [\boldsymbol{d}^\dagger_{\bar{\mathbf{q}}'_1} \otimes \boldsymbol{d}^\dagger_{\bar{\mathbf{q}}'_2}]^{\bar{\gamma}_a} \right]^{\gamma_0}_{\mu_0} |\tilde{0}\rangle \,. \qquad (9)
\end{aligned}
$$

The vacuum coefficients $Z^{n\pi_n,\gamma_n}_{(i)\ \mathbf{q}'_1\mathbf{q}'_2,\mathbf{q}_1\mathbf{q}_2}$ expression have been shown and explained in [29].

The terms $\left[\boldsymbol{b}^\dagger_{\mathbf{q}_1} \otimes \boldsymbol{d}^\dagger_{\bar{\mathbf{q}}_2}\right]^{\gamma_p}_{\mu_p}$, $\left[\boldsymbol{b}^\dagger_{\mathbf{q}_1} \otimes \boldsymbol{b}^\dagger_{\mathbf{q}_2}\right]^{\gamma_a}_{\mu_a}$ and $\left[\boldsymbol{d}^\dagger_{\bar{\mathbf{q}}_1} \otimes \boldsymbol{d}^\dagger_{\bar{\mathbf{q}}_2}\right]^{\gamma_r}_{\mu_r}$, represent the *ph*-pair operators (or quark-antiquark operators), the di-$q$ and di-$\bar{q}$ operators, respectively. Their explicit expressions have been shown in [29]. These operators are the building blocks of the RPA operators, as we will show in Section 3.2, and will be crucial for the mapping of the effective Hamiltonian, Eqs. (6)-(7), to a Hamiltonian that describes the propagation of the effective quarks, antiquarks, the RPA-pair and their interactions.

### *3.2. ph-, di-quark- and di-antiquark phonon operators.*

The one-phonon creation-operators of the *ph*-, di-quark- and di-antiquark-type are defined as follows

$$
\boldsymbol{\Gamma}^\dagger_{p\pi_p\gamma_p\mu_p} = \sum_{\mathbf{q}_\alpha,\mathbf{q}_\beta} \left\{ X_{\mathbf{q}_\alpha\mathbf{q}_\beta;p\pi_p\gamma_p} [\boldsymbol{b}^\dagger_{\mathbf{q}_\alpha} \otimes \boldsymbol{d}^\dagger_{\bar{\mathbf{q}}_\beta}]^{\gamma_p}_{\mu_p} - Y_{\mathbf{q}_\alpha\mathbf{q}_\beta;p\pi_p\gamma_p} (-1)^{\phi_{\gamma_p\mu_p}} [\boldsymbol{d}^{\bar{\mathbf{q}}_\beta} \otimes \boldsymbol{b}^{\mathbf{q}_\alpha}]^{\bar{\gamma}_p}_{\bar{\mu}_p} \right\} \qquad (10)
$$

$$
\boldsymbol{\Gamma}^\dagger_{a\pi_a,\gamma_a\mu_a} = \sum_{\mathbf{q}_\alpha,\mathbf{q}_\beta} \left\{ X'_{\mathbf{q}_\alpha\mathbf{q}_\beta;a\pi_a\gamma_a} [\boldsymbol{b}^\dagger_{\mathbf{q}_\alpha} \otimes \boldsymbol{b}^\dagger_{\mathbf{q}_\beta}]^{\gamma_a}_{\mu_a} - Y'_{\mathbf{q}_\alpha\mathbf{q}_\beta;a\pi_a\gamma_a} (-1)^{\phi_{\gamma_a\mu_a}} [\boldsymbol{d}^{\bar{\mathbf{q}}_\beta} \otimes \boldsymbol{d}^{\bar{\mathbf{q}}_\alpha}]^{\bar{\gamma}_a}_{\bar{\mu}_a} \right\} \qquad (11)
$$

and

$$
\boldsymbol{\Gamma}^\dagger_{r\pi_r,\gamma_r\mu_r} = \sum_{\mathbf{q}_\alpha,\mathbf{q}_\beta} \left\{ X''_{\mathbf{q}_\alpha\mathbf{q}_\beta;r\pi_r\gamma_r} [\boldsymbol{d}^\dagger_{\bar{\mathbf{q}}_\alpha} \otimes \boldsymbol{d}^\dagger_{\bar{\mathbf{q}}_\beta}]^{\gamma_r}_{\mu_r} - Y''_{\mathbf{q}_\alpha\mathbf{q}_\beta;r\pi_r\gamma_r} (-1)^{\phi_{\gamma_r\mu_r}} [\boldsymbol{b}^{\mathbf{q}_\beta} \otimes \boldsymbol{b}^{\mathbf{q}_\alpha}]^{\bar{\gamma}_r}_{\bar{\mu}_r} \right\}, \qquad (12)
$$

respectively, where the amplitudes $X$ and $Y$ are the forward- and backward-going amplitudes, in each case. The one-phonon annihilation operators $\mathbf{\Gamma}^{n\pi_n\gamma_n\mu_n} = \left(\mathbf{\Gamma}^{\dagger}_{n\pi_n,\gamma_n\mu_n}\right)^{\dagger}$ satisfy $\Gamma^{n\pi_n\gamma_n\mu_n}|RPA\rangle = 0$, for $n = p, a, r$. The phase $(-1)^{\phi_{\gamma_n\mu_n}} = (-1)^{J_n - M_{J_n}}(-1)^{\chi_{C_n}}(-1)^{T_n - M_{T_n}}$ in each phonon operator $n = p, a, r$ accounts for the scalar couplings in the vacuum, Eq. (9).

In order to explore the effective QCD Hamiltonian by means of the RPA method within colorless, like in [9], and colored subspaces like those spanned by the di-$q$ and di-$\bar{q}$ phonon operators, it is important to distinguish the interactions, see Figure 1, that will contribute to the $ph$ subspace and those to the di-$q$(di-$\bar{q}$) subspaces of RPA equation,

$$\begin{aligned} \langle RPA| \left[\hat{\Gamma}^{n'\pi_n\gamma_n\mu_n}, \left[\boldsymbol{H}^{QCD}_{eff}, \Gamma^{\dagger}_{n\pi_n\gamma_n\mu_n}\right]\right] |RPA\rangle \\ = \omega_{n\pi_n\gamma_n}\delta_{n,n'}, \end{aligned} \qquad (13)$$

with $\Gamma^{\dagger}_{n\pi_n\gamma_n\mu_n}$ the collective RPA operators and $\omega_{n\pi_n\gamma_n}$ the eigenvalues for the corresponding $ph$- di-$q$- di-$\bar{q}$-subspace. The latter implies four color subspaces of diagonalization *i.e.*, $C_p = (00), (11)$ and $C_a = (20), (01)$, since the corresponding diagonalizations in the subspaces $C_r = (02)$ and $C_r = (10)$ are symmetric with respect to $C_a = (20)$ and $C_a = (01)$ respectively.

### *3.3. Baryon-like operators and representations.*

We now introduce the baryon representations obtained from the effective quarks and the phonon operators described above.

The simplest baryon-like operators in this scheme are given by

$$\boldsymbol{B}^{\dagger}_{B_1\pi_{B_1},\gamma_{B_1}\mu_{B_1}} = \left[\boldsymbol{b}^{\dagger}_{\mathbf{q}_1} \otimes \boldsymbol{b}^{\dagger}_{\mathbf{q}_2} \otimes \boldsymbol{b}^{\dagger}_{\mathbf{q}_3}\right]^{\gamma_{B_1}}_{\mu_{B_1}}, \qquad (14)$$

where $\pi_{B_1} = \prod_i \pi_{q_i}$ and $\gamma_{B_1} = \{J_{B_1}, (Y_{B_1} T_{B_1}), C_{B_1}\}$. These baryon-like operators do not account for any ground state correlations.

The first baryon-like operators that account for ground state correlations are

$$\boldsymbol{B}^{\dagger}_{B_2\pi_{B_2},\gamma_{B_2}\mu_{B_2}} = \left[\boldsymbol{B}^{\dagger}_{B_1\pi_{B_1},\gamma_{B_1}} \otimes \mathbf{\Gamma}^{\dagger}_{p\pi_p,\gamma_p}\right]^{\gamma_{B_2}}_{\mu_{B_2}}. \qquad (15)$$

Clearly the number of baryon-like representations for the $\boldsymbol{B}^{\dagger}_{B_2}$ operator increases considerable respect to the $\boldsymbol{B}^{\dagger}_{B_1}$ operator, due to all the possibilities to couple to baryon quantum numbers. In Section 4.1, we present the results for the lowest baryon-like states and discuss the more important features.

The other baryon-like operator that accounts for ground state correlations is

$$\boldsymbol{B}^{\dagger}_{B_3\pi_{B_3},\gamma_{B_3}\mu_{B_3}} = \left[\left[\boldsymbol{b}^{\dagger}_{\mathbf{q}_1} \otimes \left[\boldsymbol{b}^{\dagger}_{\mathbf{q}_2} \otimes \boldsymbol{d}^{\dagger}_{\bar{\mathbf{q}}_3}\right]^{\gamma_0}\right]_{\mathbf{q}_1} \otimes \mathbf{\Gamma}^{\dagger}_{a,\pi_a\gamma_a}\right]^{\gamma_{B_3}}_{\mu_{B_3}} \qquad (16)$$

where this operator can be seen as a quark in the presence of a simple scalar $ph$-pair and a di-$q$ phonon. Therefore, in order to describe a positive parity baryon-like state, the quark denoted by $\mathbf{q}_1$ has the same parity as the di-$q$ phonon operator $\pi_{q_1} = \pi_a$.

With the description of the baryon-like operators in terms of the effective degrees, we can analyze the effective Hamiltonian, searching for those interactions that will be relevant for the description of a more general baryon-like state.

IOP Conf. Series: Journal of Physics: Conf. Series **1078** (2018) 012021 doi:10.1088/1742-6596/1078/1/012021

## 4. The mapped Hamiltonian.

In a first stage the RPA method leaves out the Hamiltonian terms $H_{31}$ and $H_{13}$, *i.e.* second and third terms in Eq. (7), corresponding to three creation and one annihilation operators and their hermitian conjugate terms. These terms are taken into account by the transformed Hamiltonian of Eq. (17). Once the RPA method is implemented and their solutions known, the effective QCD Hamiltonian can be mapped onto a Hamiltonian that describes the propagation of the original fermionic particles (quarks and antiquarks) as well as the boson-like particles (*ph*, di-$q$ and di-$\bar{q}$ phonons) and the interaction between them. The procedure to get the vertex functions ($\Lambda^{(i)}_{n\pi\gamma;\mathbf{qq'}}$) is rather straightforward: one needs to return to the effective Hamiltonian, Eqs. (6)-(8), and replace the simple *ph*-pair, di-$q$ and di-$\bar{q}$ operators by linear combinations of phonon operators. In Figure 2, we illustrate the mapping of two $H_{31}$ terms by replacing a *ph*-pair and a di-$q$ by a linear combination of phonon operators. The latter is done by inverting Eqs. (10),

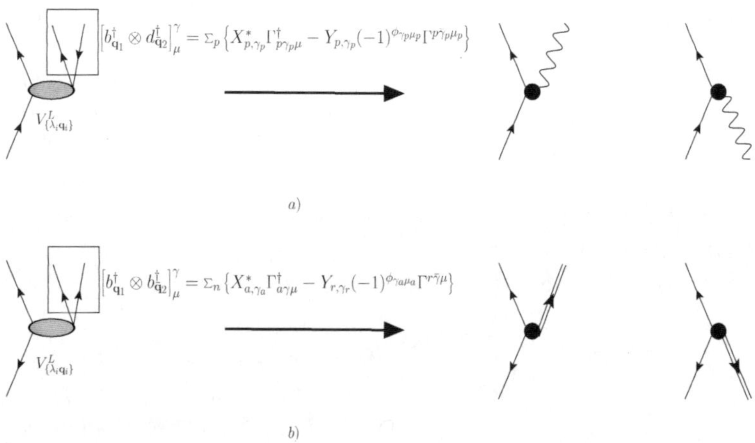


**Figure 2.** Schematic representation for the mapping of the $H_{31}$ terms onto their bosonized representation. a) the creation of a *ph*-pair is expressed in terms of *ph*-RPA phonons, b) the creation of a di-$q$ is replaced by its phonon counterpart.

(11) and (12), where the pseudo-norm $X^2 - Y^2 = 1$ is satisfied by the forward and backward amplitudes of each phonon operator.

The general form of the mapped effective Hamiltonian is given by

$$
\begin{aligned}
H^{QCD}_{eff,map} &= \sum_{\mathbf{q}\mu_q} \varepsilon_{\mathbf{q}} \left( \boldsymbol{b}^{\dagger}_{\mathbf{q}\mu_q} \boldsymbol{b}^{\mathbf{q}\mu_q} + \boldsymbol{d}^{\dagger}_{\bar{\mathbf{q}}\mu_q} \boldsymbol{d}^{\bar{\mathbf{q}}\mu_q} \right) + \sum_{n} \sum_{\pi_n \gamma_n \mu_n} \omega_{n\pi_n\gamma_n} \boldsymbol{\Gamma}^{\dagger}_{n\pi_n,\gamma_n\mu_n} \boldsymbol{\Gamma}^{n\pi_n,\gamma_n\mu_n} \\
&+ \sum_{n\pi\gamma} \sum_{\mathbf{qq'}} \Lambda^{(1)}_{n\pi\gamma;\mathbf{qq'}} \sum_{\mu} \frac{(-1)^{\gamma-\mu}}{\sqrt{\dim(\gamma)}} \left\{ \boldsymbol{\Gamma}^{\dagger}_{n=p\pi,\gamma\mu} \left( \left[ \boldsymbol{b}^{\dagger}_{\mathbf{q}} \otimes \boldsymbol{b}_{\bar{\mathbf{q}}'} \right]^{\bar{\gamma}}_{\bar{\mu}} + (-1)^{\bar{\gamma}_{\mathbf{q}'}+\gamma_{\mathbf{q}}-\bar{\gamma}} \left[ \boldsymbol{d}^{\dagger}_{\bar{\mathbf{q}}'} \otimes \boldsymbol{d}_{\mathbf{q}} \right]^{\bar{\gamma}}_{\bar{\mu}} \right) + h.c. \right\} \\
&+ \sum_{n\pi\gamma} \sum_{\mathbf{qq'}} \Lambda^{(2)}_{n\pi\gamma;\mathbf{qq'}} \sum_{\mu} \frac{(-1)^{\gamma-\mu}}{\sqrt{\dim(\gamma)}} \left\{ \left( \boldsymbol{\Gamma}^{\dagger}_{n=a\pi,\gamma\mu} - (-1)^{\gamma-\mu} \boldsymbol{\Gamma}^{n=r\pi,\bar{\gamma}\bar{\mu}} \right) \left[ \boldsymbol{d}^{\dagger}_{\bar{\mathbf{q}}} \otimes \boldsymbol{b}_{\bar{\mathbf{q}}'} \right]^{\bar{\gamma}}_{\bar{\mu}} + h.c. \right\} \\
&+ \sum_{n\pi\gamma} \sum_{\mathbf{qq'}} \Lambda^{(3)}_{n\pi\gamma;\mathbf{qq'}} \sum_{\mu} \frac{(-1)^{\gamma-\mu}}{\sqrt{\dim(\gamma)}} \left\{ \left( \boldsymbol{\Gamma}^{\dagger}_{n=a\pi,\gamma\mu} - (-1)^{\gamma-\mu} \boldsymbol{\Gamma}^{n=r\pi,\bar{\gamma}\bar{\mu}} \right) \left[ \boldsymbol{b}^{\dagger}_{\mathbf{q}} \otimes \boldsymbol{d}_{\mathbf{q}'} \right]^{\bar{\gamma}}_{\bar{\mu}} + h.c. \right\} . \qquad (17)
\end{aligned}
$$

The schematic representation of the matrix elements of the mapped Hamiltonian in the baryon-like states basis spanned by the operators given in Eqs. (14)-(16),

$$
|B\rangle = b_1|B_1\rangle + b_2|B_2\rangle + b_3|B_3\rangle \ , \qquad (18)
$$

are shown in Figure 3.

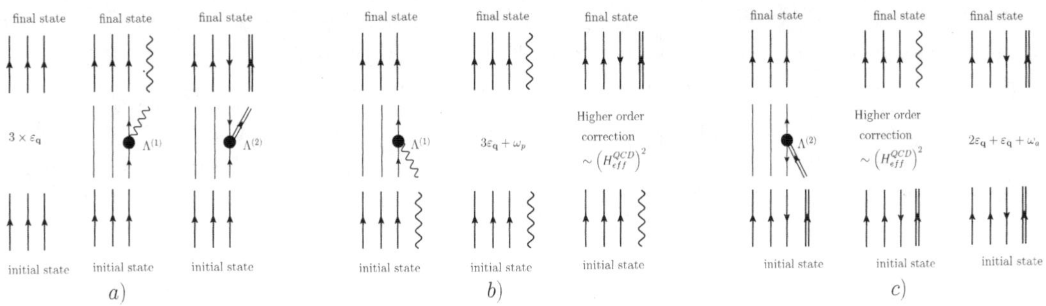

**Figure 3.** Schematic representation of the matrix elements of the mapped Hamiltonian Eq. (17) in the baryon-like basis, Eqs. (14)-(16).

*4.1. Simple calculation and results.*

In order to get some insights about the baryon spectrum by using the mapped Hamiltonian and the effective degrees of freedom described above, we focused on the description of the low energy baryons, *i.e.* $p, n$ and $\Delta$ states and their quantum numbers $T_B\left(J_B^{\pi_B}\right) = \frac{1}{2}\left(\frac{1}{2}^+\right), \frac{3}{2}\left(\frac{3}{2}^+\right)$ and $C_B = (00)$. For such purpose, we have considered the radial and angular excitations up to $N = 1$ and $l = 1$ and restricted to single quark states to total spin $J_q = \frac{1}{2}$.

In the case of the $B_2$ operator, Eq. (15), the baryon-like irreps related to the nucleon and $\Delta$ states, *i.e.*, $J_{B_2}^{\pi B_2} = \frac{1}{2}^+$ and $\frac{3}{2}^+$, $(Y_{B_2} T_{B_2}) = \left(1, \frac{1}{2}\right)$ and $\left(1, \frac{3}{2}\right)$ and $C_{B_2} = (00)$, the positive parity presents a particular scenario. Since the lowest *ph*-RPA solution corresponds to that given by the pion-like one, the $B_1$ baryon-like operator in Eq. (15), must have negative parity. Thus the negative parity coming from the $B_1$ baryon-like operator is due to a radial excitation of one the effective quarks.

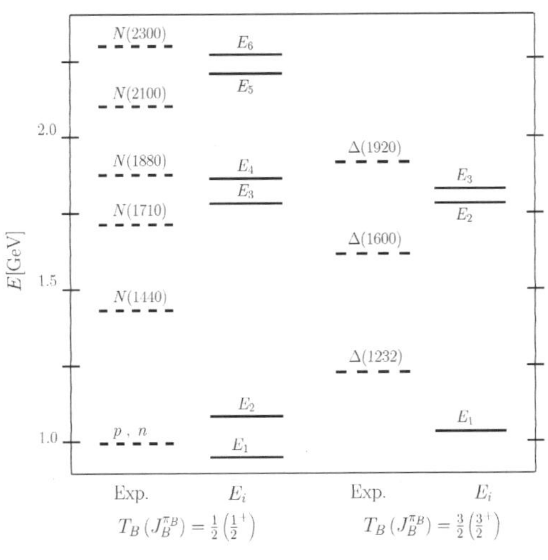

**Figure 4.** Comparison between calculated energies $E_i$ of baryon-like states (solid lines) with experimental values (dashed-lines) taken from [30].

For the $B_3$ baryon-like operator, Eq. (16), the lowest di-$q$ phonon solutions corresponds to $T_a\left(J_a^{\pi}\right) = 0\left(0^+\right)$ [29], which is the lowest RPA solutions within the SU(3) color subspace

(01), needed in order to satisfy the colorless condition of a physical state. Therefore, the $B_3$ baryon-like operators contribute only to the nucleon description but not to the $\Delta$-baryon.

For the spectrum shown in Figure 4, we have used effective vertex-coefficients, $\Lambda^{(i)}_{n\pi\gamma;\mathbf{qq}'} \rightarrow \Lambda^{(i)}$, and work in the minimal space of quark(antiquark) configurations. The latter translated in the use of three interaction parameters $\Lambda^{(1)}, \Lambda^{(2)}$ and $\Lambda^{(3)}$. The first two were used to calculate the nucleon spectrum up to 2.5 GeV and the third used to calculate the $\Delta$ spectrum up to 2 GeV. Considering the simplicity of the calculation, the spectrum resulted encouraging, since several features described by the experimental data could be observed in our results. One of them is the number of states obtained at about 2.5GeV, *i.e.*, six nucleon-like solutions and three $\Delta$-like solutions. The energy range for these solutions were also satisfactory. However, the energy gaps between the experimental states were not reproduced exactly, but from our experience with the implementation of the many-body methods, we expect that a larger quark(antiquark) configurational space generates more and more collective RPA-solution, due to the larger number of such pairs in the vacuum of the model, and with it lower energy phonon states. The latter should provide more notorious energy gaps, *e.g.* between the N(1710) and N(1880).

Within this calculation, the Roper resonance (N(1440)) was not well reproduced. However, that was something expectable, since it corresponds to a radial excitation of the nucleon and its low energy is associated with a meson cloud that shields the quark core [6]. In our calculation, we did not considered radial and angular excitations higher than $N \leq 1$ and $l \leq 1$, with the restriction of total single quark spin $J_q = \frac{1}{2}$. The first radial excitation corresponding to a positive parity involves $N = 2$ which of course should improve the description of the Roper resonance and well as the implementation of a larger configurational space, that will generate a more collective *ph*-phonon state. The *ph*-phonon in Eq. (15) plays the role of a meson-like contribution to the baryon-like description.

## 5. Conclusions.

We have presented a description of baryon-like states by identifying effective degrees of freedom within an effective QCD Hamiltonian. The main conjecture of the model is that colored pairs may contribute to the structure of the vacuum, describing part of the ground state correlations. We started from the QCD Hamiltonian within the Coulomb gauge and restricted to the pure quark sector with a confining interaction. We identified effective quarks and antiquarks which led to a coupling scheme of effective degrees of freedom. We applied the RPA method to partially diagonalized the effective Hamiltonian within collective pairs subspaces. We went further and showed how to incorporate those interactions that were left out in the RPA method, *i.e.* the $H_{31}$ and $H_{13}$ terms, and from them constructed a mapped Hamiltonian that described the interaction between effective quark(antiquarks) and the RPA-phonon states. Finally, we constructed the baryon-like basis, beyond the common three quarks description, and diagonlize the mapped Hamiltonian in a reduced configurational space. The results are quite encouraging, since they show that the use of non-perturbative many-body techniques are powerful enough to simplify the task of treating many quark and antiquark configurations within the real QCD, as needed for the microscopic interpretation of baryon states at low energy.

## Acknowledgements

One of the authors (T.Y-M) thanks the National Research Council of Argentina (CONICET) for a post-doctoral scholarships. (O.C.) is a member of the scientific career of the CONICET. D.A.A-Q acknowledges financial help from DGAPA-PAPIIT (IN100418) and from CONACYT. (P.O.H) acknowledges financial help from DGAPA-PAPIIT (IN100418) and from CONACYT (Mexico, grant 251817). This work has been supported financially by the CONICET (PIP-282) and by the ANPCYT of Argentina.

## References

[1] P. de Forcrand and M. DElia, Phys. Rev. Lett. **82**, (1999) 4582.
[2] W. Bietenholz *et al* (QCDSF-UKQCD Collaboration), Phys. Rev. D **84**, (2011) 054509.
[3] J. J. Dudek *et al.*, Phys. Rev. D **83**, (2011) 111502(R).
[4] R. G. Edwards, J. J. Dudek, D. G. Richards, and S. J. Wallace, Phys. Rev. D **84**, (2011) 074508 .
[5] G. Eichmann, H. Sanchis-Alepuz, R. Williams, R. Alkofer and C. S. Fischer, Prog. Part. Nucl. Phys. **91** (2016) 1.
[6] N. Suzuki, B. Juliá-Díaz, H. Kamano, T.-S. H. Lee, A. Matsuyama and T. Sato. Phys. Rev. Lett. **104**, (2010) 042302.
[7] A. Szczepaniak, E. S. Swanson, C. R. Ji and S. R. Cotanch, Phys. Rev. Lett. **76**, (1996) 2011.
[8] F. J. Llanes-Estrada and S. R. Cotanch, Nucl. Phys. A **697**, (2002) 303.
[9] D. A. Amor-Quiroz, T. Yépez-Martínez, P. O. Hess, O. Civitarese and A. Weber, Int. J. Mod. Phys. E **26**, (2017) 1750082.
[10] N. H. Christ and T. D. Lee, Phys. Rev. D **22**, (1980) 939.
[11] H. Reinhardt *et al.*, arXiv:1706.02702v1 [hep-th].
[12] A. P. Szczepaniak and E. Swanson, Phys. Rev. D **65**, (2001) 025012.
[13] C. Feuchter and H. Reinhardt, Phys. Rev. D **70**, (2004) 105021.
[14] D. R. Campagnari and H. Reinhardt, Phys. Rev. D **82**, (2010) 105021.
[15] M. Pak and H. Reinhardt, Phys. Rev. D **88**, (2013) 125021.
[16] P. Vastag, H. Reinhardt and D. Campagnari, Phys. Rev. D **93**, (2016) 065003.
[17] D. R. Campagnari, E. Ebadati, H. Reinhardt and P. Vastag, Phys. Rev. D **94**, (2016) 074027.
[18] F. J. Llanes-Estrada, S. R. Cotanch, A. P. Szczepaniak and E. S. Swanson, Phys. Rev. C **70**, (2004) 035202.
[19] P. Wang, S. R. Cotanch and I. J. General, Eur. Phys. J. C **55**, (2008) 409.
[20] W. Xie, L. Q. Mo, P. Wang and S. R. Cotanch, Phys. Lett. B **725**, (2013) 148.
[21] P. Bicudo and M. Cardoso, F. J. Llanes-Estrada and T. V. Cauteren, Phys. Rev. D **94**, (2016) 054006.
[22] T. Yépez-Martínez, P. O. Hess, A. P. Szczepaniak and O. Civitarese, Phys. Rev. C **81**, (2010) 045204.
[23] P. Ring and P. Schuck, *The Nuclear Many Body Problem* (Springer, Heidelberg, 1980).
[24] Jeff Greensite and Adam P. Szczepaniak, Phys. Rev. D **91**, (2015) 034503.
[25] Jeff Greensite and Adam P. Szczepaniak, Phys. Rev. D **93**, (2016) 074506.
[26] P. Guo, A. P. Szczepaniak, G. Galata, A. Vassallo and E. Santopinto, Phys. Rev. D **78**, (2008) 056003.
[27] P. Guo, T. Yépez-Martínez and A. P. Szczepaniak, Phys. Rev. D **89**, (2014) 116005.
[28] J. Escher and J. P. Draayer, J. Math. Phys. **39**, (1998) 5123.
[29] T. Yépez-Martínez, O. Civitarese and P. O. Hess, Eur. Phys. J. A **54**, (2018) 22.
[30] Particle Data Group (C. Patrignani *et al.*), Chin. Phys. C **40**, (2016) 100001.